The Soul of Geography

Bojie Fu

The Soul of Geography

Integration is the Essence and Soul of Geography

创于1897 商務印書館 The Commercial Press

Springer

Bojie Fu
Research Center for Eco-Environmental
Sciences, Chinese Academy of Sciences
Beijing, China

ISBN 978-981-95-1061-0 ISBN 978-981-95-1062-7 (eBook)
https://doi.org/10.1007/978-981-95-1062-7

Jointly published with The Commercial Press, Ltd.
The print edition is not for sale in China (Mainland). Customers from China (Mainland) please order the print book from: The Commercial Press, Ltd.
ISBN of the Co-Publisher's edition: 978-7-100-24222-6

The original submitted manuscript has been translated into English. The translation was done using artificial intelligence. A subsequent revision was performed by the author(s) to further refine the work and to ensure that the translation is appropriate concerning content and scientific correctness. It may, however, read stylistically different from a conventional translation.

This Springer imprint is published by the registered company Springer Nature Singapore Pte Ltd.
The registered company address is: 152 Beach Road, #21-01/04 Gateway East, Singapore 189721, Singapore

Preface

Ever since I stepped into the field of geography 46 years ago, I have been constantly running and walking, without looking back or stopping. It was not until 2 months ago, when I received a visit from Mr. Li Ping, the former general manager of the Commercial Press, and Ms. Li Juan, the director of the Geography Editorial Office. They invited me to give a lecture at the Commercial Press to introduce the relationship between geography and society to the public, and also invited me to consider writing for the *Geographer's Essays* series. Their invitation made me "slow down".

Geography is a broad and profound discipline, yet public understanding often falls short of grasping its true scientific depth and scope. Looking back on decades of geographical research and teaching, I have taken a number of scattered reflections and informal writings from the past and organized them into four thematic sections: Integrated Essence, Space and Process, Systems Thinking, and Social Applications. These pieces fall somewhere between science and literature—neither strictly academic nor conventional essays. Yet in another sense, they capture the spirit, essence, and allure of geography, offering a glimpse into the insights and inner journey of a geographer.

Beijing, China
March 2024

Bojie Fu

Prologue: Geography Chose Me

The 1977 national college entrance examination was a turning point in my life. It was the most memorable exam, although not the most nerve-wracking moment. In fact, it was only 10 days before the exam that I got the qualification to apply! This opportunity came so unexpectedly, I hardly had time to think, let alone get nervous, before I walked into the exam room.

A Big Exam that Came Too Soon to Get Nervous

I grew up in a family of teachers and learned to read early. In 1963, soon after my 5th birthday, I loved leaning against the window of the classroom in my father's school. When the teacher asked questions inside, I would raise my hand and answer from outside. At first, the teacher resented my interference, but after a while, she relented and let me join the class. Thus began my academic journey—by tagging along.

In January 1974, I graduated from Xianyang No. 15 Middle School in Shaanxi Province and went to the countryside to work. Because of excellent grades and performance, I taught at a rural school, served as the Secretary of the Youth League Branch and the Deputy Secretary of the Party Branch of the Brigade, and organized political night school and poetry recitals for the educated youth. After 2 years of rural labor practice, I qualified for the last cohort of worker–peasant–soldier recommendations to university in 1976. At that time, there were three enrollment quotas in the commune where I was located. Based on my performance, I should have become a worker–peasant–soldier college student at what is now Northwest A&F University—only to have my place taken at the last moment.

Despite the disappointment, I remained true to my convictions. Even though the dream of attending university had slipped away, any chance to continue studying—regardless of the institution—was a path I could not afford to ignore. Coincidentally, I immediately received a notice from the city's recruitment office that I could enter Xianyang Machine Manufacturing School. That school, under the First Ministry of

Machinery Industry, was reputable in mechanical manufacturing. In January 1977, I enrolled in its best major at the time: machine tool electric.

To my surprise, nine months later, the chance to go to a university reappeared: a broadcast from our affiliated factory announced the reinstatement of the national college entrance examination. Like many others, I immediately prepared to apply. However, fate played yet another trick: regulations barred us, as technical school students, from registering—only fresh high school graduates and other youth could sit the exam.

Looking back, my path to university was full of twists and turns, but fortunately, in the end, the tides turned in my favor.

Just 10 days before the exam, our school suddenly received permission to let us sit as high school students, albeit with strict quotas: only five per class. Over 130 applied across the school for just 30 spots. Therefore, we had to first participate in a ranking examination organized by the school to compete for qualification to sit the college entrance examination. Fortunately, my grades ranked first. At that time, there was only one week left before the college entrance examination.

It Would be Good If One or Two Could be Admitted

I experienced the turbulent "Cultural Revolution" during my junior and senior high school years. The schooling was compressed to 2 years and 2.5 years, respectively. Especially in junior high school, apart from Chinese and mathematics, I had hardly any systematic exposure to physics, chemistry, biology, and geography. Instead, some more practical industrial and agricultural basic courses were substituted, and academic learning was essentially suspended. By the time I reached senior high school, Deng Xiaoping had returned to leadership, and the school atmosphere was slightly improved. In technical school, I focused on workshop skills far removed from core academic subjects.

In December 1977, I sat the first exam—mathematics—at Xianyang Northwest Cotton Mill Factory's middle school for workers' children. About 50 of us took it; my neighbor, an examinee 1966 graduate, asked if I'd studied the limit of a sequence—one of the foundations of advanced mathematics. We hadn't even covered trigonometry or solid geometry. I could only smile bitterly and say "no". Before the exam, the invigilator comforted us, asking us to relax and not be nervous. In his view, "it would be good if one or two of us could be admitted from our examination room".

That year, provinces set their own examination papers; Shaanxi's mathematics was tough, although by today's standards it was basic junior—high to early senior-high material. For us, without a solid grasp of fundamentals, it felt like a huge barrier. Thankfully, the Politics, Chinese, and Physics–Chemistry exams did not "torture" us quite as much. I recall the two-choice essay: "An Unforgettable Day" or "A Letter to the National Science Conference." I chose the first, writing about the nationwide publication of Volume V of *Selected Works of Mao Zedong* on April 15, 1977. In physics and chemistry exam, I solved nine of ten questions; the lone chemistry problem I couldn't answer, required the reaction formula needed for molecular-weight calculations which I did not know. I was struck by the no-shows: after the first exam, the afternoon political exam saw attendance drop by nearly one third. In my room, only about 60% of the original students completed all four papers. Perhaps noticing my ease with answers, the invigilator invited me, as our room's representative, to a feedback session organized by the Education Bureau and the Admissions Office after the final exam to discuss the papers.

Everything happened so fast that I didn't think too much about the results; when winter break arrived, I happily returned home for the New Year. Although official results were not yet released, Shaanxi Province preliminarily selected some candidates for a physical examination. Of our original 30 from Xianyang Machine Manufacturing School, 15 made that list. The outcome was, however, far beyond my expectations as, among the 15, I was the only one fortunate enough to pass entrance to university.

Racing Against Time

At the end of February 1978, without much time to say farewell to my classmates at Xianyang Machine Manufacturing School, I withdrew to enroll at Shaanxi Normal University. In the eyes of my classmates, this university was represented an under-achievement. Influenced by national needs, I had listed Tsinghua University's Automatic Control as my first choice, Xi'an Jiaotong University's Radio Engineering second, and Northwestern Polytechnical University third. Nevertheless, my admission letter read "Geography". I suspected a clerical error: they'd mistaken "Physics (Wuli)" for "Geography (Dili)." We joked that geography was so easy "if you know 500 Chinese characters, you can study it—why do you need to study geography for four years?" In fact, of the 70 students admitted to the Geography Department of Shaanxi Normal University at that time, all but 2 were transferred to this major. Some students even went to the provincial admissions office to argue, unwilling to accept the arrangement. In the final analysis, it was because they lacked even a basic understanding of this major.

With prejudice against both geography and normal universities, I started my university life. What impressed me most was the cafeteria. Having experienced natural disasters and the difficult period of going to the countryside, we young people were finally treated to abundant food; breakfast even included pastries—once a luxury. As a comprehensive university with both arts and sciences, I was quickly attracted by its strong academic, cultural, and artistic atmosphere. The library's retro architecture was beautiful. I remember there were two huge cedars at its entrance, standing there since the university was founded.

Not long after we began, the National Science Congress convened. This "spring of science" profoundly influenced our generation. In particular, the paper in *People's Literature* on the *Goldbach Conjecture*—telling the stirring story of mathematician Chen Jingrun—was spread nationwide. It showed us the value of scientific inquiry and intellectual dedication. Our campus overflowed with thirst for knowledge and active exploration. Having experienced an era of intellectual and spiritual deprivation, we unleashed a tremendous energy. Because we had once been deprived, we felt we must race against time to catch up. Everyone was studying hard, spending time in the library, reading literature, and continuing to study under the streetlights after the dormitory lights were turned off. This would seem unimaginable to today's young students.

It was at this time that I developed the habit of keeping reading and research notebooks, some of which I still own. For me, although I didn't have a clear goal to devote myself to science at that time, my desire for knowledge had never been so strong.

Field Practice Experience Benefits Me for Life

Starting from my sophomore year, I gradually developed an interest in geography. I have always been the kind of student that wanted to balance both arts and sciences, and geography is a cross-disciplinary subject that combines natural sciences and social sciences. It involves a wide range of aspects—geomorphology, soil geography, biogeography, human geography, economic geography, meteorology, climatology, and many more. I eventually realized that the "integrated" essence of geography was very much aligned with my interests.

By the second semester of junior year, the university invited a Chinese-American professor from California State University to teach three specialized courses. His arrival introduced us to urban geography, population geography, behavioral geography, and other new concepts, instantly broadening our research horizons. Such a complex subject as geography requires complex methods, and science had not yet fully conquered the problems of complex systems; thus, a vast exploratory space remained. Aside from reinforcing fundamentals, a hallmark of our geography program was fieldwork. Even today, I tell my students that the extensive fieldwork I did in university has benefited me for life.

In October 1978, just 1 month into the second semester of the undergraduate period, we conducted a 1-month field practice in measurement and cartography, mapping a 1:2000 topographic map for a mining area. Thereafter, every semester's holiday included mandatory fieldwork—2 weeks at minimum, up to a month. This greatly differs from most current geography training.

During my postgraduate period, our field skills were further honed. I traveled all over the country—plateaus, glaciers, deserts, islands—learning and recording each region's soils, vegetation communities, hydrology, and landforms in detail. Each time, we set off with our bedding and slept in abandoned factories or warehouses at night. Despite the very tough conditions, we always enjoyed it. In addition to improving professional skills, it also created a strong emotional connection among classmates. I remember when I was doing geological fieldwork in Hanzhong during my undergraduate studies, in addition to Shaanxi Normal University, there were also teams from Xi'an College of Geology, Northwestern University, and Gansu Normal University. Initially, we felt competitive and argued in the communal area. By the end, however, we lived and studied together, organized entertainment, and created wonderful memories.

In undergrad, the most influential teacher for me was Prof. Liu Yinhan, teacher of Chinese Physical Geography and later my graduate advisor. Professor Liu's classes were distinctive: He always carried two thick volumes: *Soils of China* and *Vegetation of China.* In that pre-projector era, he passed around the pictures and diagrams in the books for us to browse. Professor Liu was very strict. During our comprehensive internship in the senior year of 1981, we took a nearly 20-hour train journey from Xi'an to Ankang, but he insisted we begin studying the physical geography of Ankang that very night. Some unaware classmates had bought movie tickets in advance to relax. As a member of the study committee, I begged Prof. Liu to relent but he was angry and severely criticized our behavior. He left us saying: "If you study well, there will be small cars inviting you to watch 'Caravan' (a movie)".

Finding Research Direction

In my fourth year, I graduated top of my class with straight A's. In January 1982, after a fierce master's entrance examination, I was admitted as a graduate student in the Department of Geography at Shaanxi Normal University, specializing in regional geography, with a research focus on integrated physical geography. After 4 years of undergraduate study, I recognized the breadth and depth of geography, but faced the critical question: what research direction to pursue?

Geography's soul is integration; the landscape is the basic unit of human–natural interaction—the smallest unit for integrated geographic research. Centering on the term "landscape", I enthusiastically collected and reviewed the relevant literature. Among the wealth of readings, the emerging field of "landscape ecology" caught my eye. It was a new interdisciplinary field that developed in the early 1980s, representing the intersection of geography and ecology. The first international conference for landscape ecology had been held in the Netherlands in 1981. Based on my literature review and analysis, as a new graduate student, I prepared a paper in 1982 and in 1983, I published my first academic paper—China's first article in landscape ecology: "*A New Frontier in Geography—Landscape Ecology*". I adopted it as my research direction. Ever since then I have been working in and exploring this field.

During my master's degree, I published seven academic papers in core domestic journals such as *Acta Ecologica Sinica* and *Geographical Science.* After graduation, I was invited to stay on as faculty. However, in the spring of 1986, deciding to continue my studies, I sent my published articles to renowned geographer Prof. Lin Chao of the Department of Geography at Peking University. He secured me an exemption from the entrance exam. After English and politics examinations, I entered Peking University for doctoral studies. With his recommendation, in 1987, I became among the first cohort of Chinese doctoral students jointly trained abroad, heading to the

University of Stirling in the UK to engage with international frontier topics. In my entire academic journey, if my undergraduate and master's advisor Prof. Liu Yinhan helped me establish strong local foundations, Prof. Lin Chao opened my international perspective—a philosophy I now strive to pass on to my students.

Conducting Landscape Ecology Research

After obtaining my doctorate in 1989, I was recruited by the Research Center for Eco-Environmental Science, Chinese Academy of Sciences. At that time, I realized that landscape ecology needed to strengthen the coupling of patterns and processes, which requires the integration of the "pattern" emphasis of geography and the "process" emphasis of ecology. Yet ideas alone cannot be translated directly into practice. Early in my career, funding was my main hurdle: the institute gave just 1000 yuan startup fund; the Ministry of Education approved a 20,000 yuan returnee fund. After buying a computer, I couldn't even afford a printer. The following year, although the National Natural Science Foundation of China approved a grant of 30,000 yuan, this was like a drop in the bucket for process-monitoring research. Due to insufficient research funding, many experiments and observations could not be carried out, and many ideas could not be realized, which was quite frustrating at the time.

Despite difficulties, I sought involvement in China's eco-environmental early-warning research. When idle, one must find work oneself. In 1992, unable to fully express myself at home, I went to Belgium for postdoctoral research, systematically studying landscape ecology principles and methods, and delving into international frontier issues. Upon returning in 1994, my proposals such as "Land-Use Structure and Ecological Processes in the Loess Hilly-Gully Region" secured the National Nature Science Fund for General Program and Distinguished Young Scholars, also the EU international cooperation fund. With that initial funding, I was finally realizing my ideas and truly began practicing landscape ecology research.

Through field-site selection and investigation, I chose Yangjuangou in Baota District, Yan'an City, as our base for studying the landscape pattern and ecological processes of the Loess Plateau. We carried out long-term observational research in this typical loess hilly and gully area of northern Shaanxi. In the process, I started to form a research team.

As Prof. Huang Bingwei, a famous geographer in China, said, integrated geographic research requires a group of people to work regularly, to understand each other, and gradually become a cohesive collective. This collective should consist of about 10 people—too few limits breadth; too many fragments into cliques. Guided by this vision, we gradually assembled a multidisciplinary team of around ten researchers, spanning generations and bringing expertise in geography, ecology, hydrology, soil science, remote sensing, and geographic information systems to pursue a shared research agenda.

From the Loess Plateau to the World

Starting from 1996, using the Loess Plateau as our monitoring base, I led my team and students in long-term field monitoring and multi-scale observation, experimental and model simulation research. We took the lead in integrating scale, pattern, and process, systematically revealing how land-use patterns in the loess hilly and gully area influence soil moisture, nutrients, and erosion. We developed a spatial distribution model of soil moisture in loess hillslopes and proposed optimal land-use structures for loess hillslopes and small watersheds. These pioneering ideas and methods deepened integrated physical geography's pattern and classification research into process studies, earning the 2005 National Natural Science Award (Second Prize).

After 2008, I directed my team to ecosystem services research, extending research on landscape patterns and ecological processes to ecosystem processes and services at various scales—the "new integration" of geography. We developed methods for trade-off analysis of ecosystem services, revealing the trade-off relationship and its spatiotemporal variation between carbon sequestration and water yield services during vegetation restoration; determined the water resource carrying capacity threshold for vegetation restoration in the Loess Plateau, and found that current revegetation is approaching sustainable water resource limits; developed a spatially explicit assessment and optimization tool for regional ecosystem services, and proposed land use allocation schemes oriented to multi-objective optimization of ecosystem services. These works led ecosystem services research into processes, mechanisms, and integrated optimization, providing scientific support for ecological restoration in the Loess Plateau. In 2020, our research again won the National Natural Science Award (Second Prize).

As human impacts on the Earth system intensify, global change and sustainable development have become key themes in geographical research. To address these new challenges, in recent years, I have led my team to focus on coupled human and natural systems and sustainable development research. Building on earlier work, we proposed a framework of "Pattern–Process–Service–Sustainability", studying coupling mechanisms and optimized regulation of human–natural systems; for global SDG implementation, we proposed a "Classification–Coordination–Collaboration" systemic approach; we extended ecosystem services research to the global scale, analyzing the impacts of global change on major services and proposing comprehensive assessment methods and management strategies; we initiated and lead the Global Dryland Ecosystem Program, aiming to provide scientific bases for ecological protection, rational resource use, and sustainable development at regional, national, and global scales.

Reflecting on over 40 years of geographical study and research—from prejudice to passion, from ignorance to innovation—my journey has spanned land use types and evaluation research on the Loess Plateau, landscape patterns, and ecological processes studies, ecosystem processes and services research, to present coupled human and natural system and sustainable development research. I have persisted in one research direction, constantly delving deeper, sticking to one region and field, gradually expanding. The journey of scientific innovation is always on the road.

Originally published in Chinese in *China Science Daily*

June 2018, abridged and revised.

Fu Bojie (assisted by Hu Minqi in organizing)

Contents

About the Author

Bojie Fu distinguished professor at the Research Center for Eco-Environmental Sciences, Chinese Academy of Sciences, is an academician of the Chinese Academy of Sciences, a fellow of the Academy of Sciences for Developing World, an international honorary member of American Academy of Arts and Sciences, a corresponding fellow of the Royal Society Edinburgh, and an honorary fellow of the Royal Geographical Society. He serves as vice president of the International Geographical Union and chairman of the supervisory board of the Geographical Society of China, and he was previously president of the Geographical Society of China.

Integrated Essence

Geography is a comprehensive discipline that spans the natural sciences, social sciences, and technical sciences. Integration is the essence and soul of geography. Geography's initial comprehensiveness stemmed from its diversity; however, with the deepening development of its subfields, geography has increasingly become hollowed out. To prevent this, it is essential to strengthen foundational research in geographic theory and methods, enhance the study of human-natural interactions and their coupling mechanisms, and emphasize the integrated essence of geography.

Geography: From Knowledge, Science, to Decision Making Support

Geography constitutes an important part of humanity's repository of knowledge, aiming to "explore natural laws and reveal humanistic essence". It is an ancient yet modern science, as early as on the *Book of Changes* (*Zhou Yi*), it is stated, "looking up to observe astronomy, looking down to observe geography". In modern geography ("dili" in Chinese), "di" refers to the Earth, or the Earth's surface, or to a region, while "li" denotes the principles and laws of things, or the intrinsic connections governing phenomena. "dili", then, addresses the spatial distribution, temporal evolution, and interactive laws of phenomena or entities on the Earth's surface. Geography as a discipline investigates the spatial distribution patterns, temporal evolution processes, and regional characteristics of geographic elements or geographic complexes. Geographic elements typically include water, soil, atmosphere, biota, and humans. A geographic complex comprises these geographic elements. In nature, an ecosystem or a natural zone can be viewed as a geographic complex; in human society, a city—or even a city block—can serve as a geographic complex. The study of spatial distribution patterns—from Xishuangbanna in the south to Mohe in the north—encompasses a latitudinal span from the tropics to the frigid zone. Such distribution of

B. Fu, *The Soul of Geography*, https://doi.org/10.1007/978-981-95-1062-7_1

natural zones exemplifies spatial differentiation of geographic complexes. The study of temporal evolution primarily concerns historical periods and includes aspects of geological history—for example, in Quaternary studies we analyze geographic processes from the past to the present, and then predict future changes, which constitutes temporal evolution. Regional characteristics of geographic elements and complexes also fall within geography's purview. For instance, in researching coordinated development in the Beijing–Tianjin–Hebei region, its range of regional features are analyzed, considering industrial complementarity, functional roles, transportation network composition, and so on, with the aim of elucidating intercity interactions and thus serve regional coordination and sustainable development.

Through long-term development, particularly in recent decades, geographic theories, technologies, and methods have continuously innovated, and research paradigms are quietly shifting. Geography has clearly embraced the modern science and technology system. With its integrative and systematic perspective, geography is at the forefront of contemporary Earth sciences, which are transitioning into a new phase focused on Earth system science. Meanwhile, the major societal needs bred by the increasingly complex human–natural interactions provide new space for geography to address the many complex issues of resources, environment, and development. Today, geography faces favorable opportunities and is in a historical process of transforming into geographical science, while it continues to develop and improve in serving government decision-making and national needs.

The Splendid Transformation from Geography to Geographical Science

Geography's research goals extend beyond explaining the past—they crucially serve the present and forecast the future. Traditionally qualitative, geography is undergoing a magnificent transformation into a science with unique research methods and quantification, which is mainly reflected in the following three aspects.

First, transformation of research methods. Traditional geographic methods include surveying, observation, recording, mapping, zoning, and planning. Early zonings guided macro-level decision support but lacked detailed micro-level information. Modern geography retains these strengths while enhancing fieldwork and observation, and increasingly applies spatial statistics, remote sensing, geographic information systems (GIS), and other technologies to build models and decision-support systems that directly serve management and decision-making.

Second, the research technology of geography has moved from conceptual models to quantitative expressions. From early geographical zonal concepts, Köppen climate classification to current climate system models, from simple spatial overlays in GIS to multi-sphere, multi-element coupling in Earth-system models—the simulation accuracy of geographic models is ever improving. GIS has evolved from computational cartography and spatial analysis into digital-Earth systems with massive multiresolution data and 3D visualization. Remote sensing has progressed from aerial photography to multi-satellite, multiresolution, all-weather, full-spectrum, multi-element Earth observation, greatly enhancing temporal and spatial resolution. Global positioning has advanced from compass to satellite networks to mobile-network-based location services. As monitoring systems for the Earth's surface mature—ranging from spaceborne and airborne remote sensing to underground detection and multi-element surface observations (soil, vegetation, water)—refined and multi-scale data abound, deepening our understanding of geographic processes and mechanisms. Geography research has moved from statistical models to simulation modeling, from linear analyses to nonlinear mathematical statistics, and from single models to model systems: predictive, multi-sphere, multi-element coupled Earth-system models are now possible.

Third, big data, visualization, and virtual reality provide essential tools for tackling complex geographic problems. Big data can finely characterize the spatiotemporal dynamics of socioeconomic phenomena. For example, in the analysis of human phenomena, hotspots can be identified through mapping population flows, thereby providing a scientific basis for the public facility layouts and transportation networks; integrating indoor simulations and outdoor observations—combining observed geographic process data with digital rainfall, vegetation cover, and urban traffic network layouts—enables intelligent, visual simulations on computers that support decision-making. Early land-cover, land-use, and soil mapping relied on manual field surveys and hand-drawn maps using sample plots for different mapping units. Now, remote sensing image classification and GIS integration of other spatial data enable the study of human-natural interactions—such as land use and land cover change, urbanization processes and drivers—and serve biodiversity conservation, urban–rural planning, and global change responses.

Evolution of Geographical Research Themes

Geographical research themes continue to evolve—from multifaceted approaches to more systematic ones—emphasizing the study of the Earth's surface systems, particularly terrestrial surfaces, through a systems perspective and the analytical tools of geographical science to address major challenges facing human society. The rapidly changing Earth's surface provides important opportunities for the development of geographical science. In the UK, terrestrial surface research is a priority funded by the Natural Environment Research Council and is among the fastest-growing fields. While project scopes broaden, strong support remains for climate change and biogeochemical cycle research. In the US, funded research emphasizes land use, climate, water resources, and human activities, stressing multidisciplinary integration. Large-scale studies of surface-change processes that combine climate, land use, ecosystems, water resources, and human activity are key priorities for the National Science Foundation, which supports integrative approaches across geography, Earth sciences, life sciences, social sciences, and engineering. The National Science Foundation's support for research related to sustainable development is also increasing, and research on the coupled human and natural systems and human–natural interactions has always been a core theme. In Germany, geography projects focus on society and economy, energy, global change, resources and sustainability, Earth system science, and future cities, emphasizing the close integration of projects with Germany's actual situation to solve domestic problems.

In China, hot and frontier research areas include urbanization, climate change, environmental change, greenhouse gases, ecosystems, human–natural interactions, and land use; there is also emphasis on integrating methods such as remote sensing, GIS, spatial analysis, and digital simulation. In physical geography, water, climate change, and sustainable water and soil use are key topics. Human and economic geography focus on China's role in global economic integration. China's growing influence in the Asia-Pacific and the world also shapes research in economic, socio-cultural, and political geography. Accordingly, geography should strengthen studies of geopolitics, geo-economics, and global/regional economic integration.

A comparative analysis of geographical research in China and internationally reveals both similarities and differences. While both emphasize cities, growth, and network structures, Chinese research tends to focus more on population, land use, urbanization, and tourism. In contrast, international studies place greater emphasis on politics, health, innovation and knowledge, leisure, and gender. As a result, international work often leans toward political, cultural, and social geography, whereas Chinese research is more concentrated in physical and economic geography. Geography has now entered a new era of geographical science. Against this background, with the challenges of sustainable development as the guide, the research topics of geography will become more comprehensive and diverse, and the geographical perspective will gain broader application and recognition across disciplines.

The Evolution of Geographical Research Paradigms

Viewing spatial distribution as pattern and spatiotemporal evolution as process, geography's integrative task is to couple pattern and process and develop theories, methods, and techniques for integrated geographical research. As our understanding deepens of interactions among the lithosphere, hydrosphere, biosphere, atmosphere, and human–society sphere, geographical theories, methods, and technologies evolve rapidly, driving profound shifts in research paradigms. Currently, three paradigm stages are evident: ① knowledge description, ② coupling pattern and process, and ③ modelling the complex human and earth system. Early geography disseminated knowledge of rivers, mountains, mines, cities, railways, and so on, focusing largely on the description distributions. The current disciplinary demand is to couple patterns and processes, entering the second research paradigm of geography. The future third paradigm will simulate complex coupled human and natural systems and sustainable development systems to provide scientific bases for decision-making. Advances in GIS, remote sensing, and computing underpin these paradigm shifts. In addition, geographical research needs to grasp deeply how social and cultural phenomena couple with and drive surface environmental changes, and future geographical research will inevitably be more integrated, more forward-looking, and more quantitative.

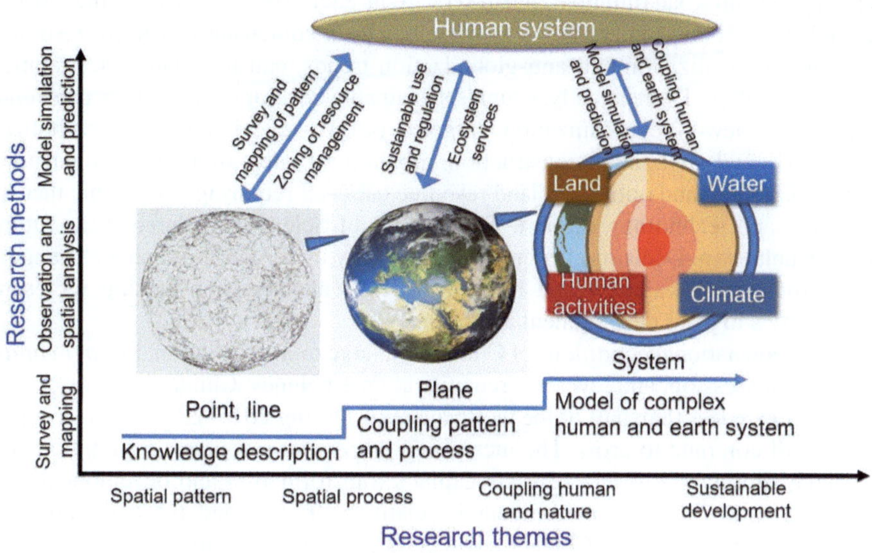

The evolution of geographical research paradigms

Globally, environment, resource utilization, and sustainability issues pose major challenges to development. The Future Earth research initiative exemplifies new-era geographical development, aiming to tightly integrate science and decision-making, and unite natural and social sciences to foster sustainable transformation worldwide. Its thematic pillars are Dynamic Planet, Global Development, and Transformations to Sustainability, with key research areas including: ① Predicting Earth-system changes—climate, carbon, biodiversity, ecosystem services, and social and economic activities; ② Exploring Earth's carrying capacity and tipping points, issuing early warnings on food, water, health, and energy under global-change; ③ Linking science knowledge to policy practice through studying potential impacts of policy, behavior, and technology choices; ④ Building high-quality multidisciplinary, multi-sector research and action collaborations to jointly design, jointly execute for application; ⑤ Supporting next-generation researchers and interdisciplinary training to enhance integrated approaches for sustainability. Anchored in moving from knowledge to science to decision-support, Future Earth offers new disciplinary space for geography's innovative development.

Geography and Decision Making

Geography plays an increasingly vital role in serving national needs and global strategies. Internationally, with the 2030 UN Sustainable Development Goals and the post-2020 Paris Agreement on climate change, China faces greater international responsibilities. Meanwhile, the global geopolitics is undergoing profound changes, such as Belt and Road Initiative, and the US "Asia-Pacific rebalancing". In the process of global economic integration, free trade has been somewhat restricted, resulting in some regionalization and anti-globalization trends, causing many uncertainties for globalization. Domestically, China's sustainable development strategy confronts challenges: new-type urbanization, targeted poverty alleviation, rural revitalization, ecological civilization construction, urban–rural integration, national territory spatial planning, and optimized land resource use—all requiring geographic theory, methods, and technology support. The essence of Science is to predict the future and the unknown. Decisions become scientific only by foreseeing change. Strengthening forecasting capacity and bridging science and decision-making represent geography's highest development aspiration.

The basic national conditions of China are a large total population, low *per capita* resource possession, and low resource utilization efficiency. China's main resources *per capita* are less than half of the world average. During 2015–2030, China's population will continue to grow. The increase in *per capita* resource consumption and resource shortages severely restricts China's transformation and development. At the same time, China's environment is relatively fragile, and the environmental pressure is great. 52% of China's territory is arid and semi-arid, with severe soil erosion, rocky desertification, and vulnerable ecosystems. Rapid urbanization in recent decades has led to a variety of compound economic–social–ecological system problems, such as urban diseases, traffic congestion, air pollution, and increased disasters. Addressing China's resource and environment challenges urgently needs scientific theories and methods—precisely where geography can shine. For example, urban planning should consider geomorphological features—preserving rivers and wetlands to ensure natural flood absorption—and intercity relations. Tianjin, once the largest northern manufacturing hub, now overlaps in its industrial layout with Beijing; coordinated development in the Beijing–Tianjin–Hebei region exemplifies geography's decision-support role. Urban planning also needs ecological perspectives—prioritizing fundamental ecological infrastructure (natural forests, wetlands) rather than merely ornamental greening, which overlooks water-conservation and pollution-control functions. Thus, macro-level geographical theory and methods are critically needed in city construction.

From China's inland–coastal and east–west disparities, to contrasts with developed Western nations, geographical research can guide decisions. China's uneven regional development—rich in western resources but facing ecological degradation and extensive development—invites reflection on the Hu Huanyong line (Heihe–Tengchong Line), which not only reflects population distribution but also natural environment characteristics. Overlaying old and new challenges intensifies China's environmental pollution: energy consumption per million-dollar GDP is about 3.4 times the world average. Western countries addressed human-waste pollution by the early twentieth century, microbial organic pollution by the 1940s, and nitrogen and phosphorus pollution by the 1980s; photochemical smog and haze were mostly resolved in the 1960s–70s. In contrast, many Chinese lakes still suffer nitrogen and phosphorus pollution—from domestic sewage and agricultural nonpoint sources—while China now faces persistent pollutants like black carbon, aerosols, greenhouse gases, organic metallic compounds, and dioxins, problems Western nations faced over a century. Thus, geography must integrate closely with decision-making to serve national needs in globalization, regional response, industrial shifts, resource flows, factor-division development, interregional competition, and geopolitical strategy.

In summary, with the advancement of geographic information technology and changes in research methods, geography in the new era is making a splendid transformation towards geographical science. Current development trends can be summarized in four aspects: ① From pattern to process research. The shift from pattern to process research confirms geography's evolution from knowledge-based to scientific. ② From element-based to system research. As the basis for the next stage of research, it is necessary not only to study the integration of natural elements and social elements, but more importantly, the integration of nature and society. ③ From theoretical to applied research. Geography aims at practical applications—advancing both theory and decision-support. ④ From knowledge creation to decision-support. The ultimate goal is for geography to progress from knowledge through science to underpin decision-making.

Originally published in Chinese in *Acta Geographica Sinica*, Issue 11, 2017, abridged and revised.

Geography: A Comprehensive Science Across Nature, Humanity, and Technology

Geography is a science that studies the relationship between humans and the Earth, characterized by its integration, interdisciplinarity, and regionality. Geography spans natural science, social science, and technical science, emphasizing interdisciplinarity from its inception. Its main branches—physical geography, human geography, and geographical technical sciences—are connected to other disciplines to varying degrees. For example, in physical geography, geomorphology intersects with geology, climatology intersects with meteorology, and biogeography intersects with biology. In human geography, population geography, urban geography, economic geography, political geography, and socio-cultural geography intersect respectively with demography, urban studies, economics, political science, and sociology. In the domain of geographical technical sciences, geographic information systems, cartography, quantitative geography, and remote sensing science intersect respectively with computer science, geodesy, mathematics, physics, and engineering.

Integration is the foundation of geography's existence—it is the soul of geography. In its early development, geography's integrative nature stemmed from the diversity of its disciplines. Professor Huang Bingwei, one of the pioneers of modern geography in China, proposed that integration is the basis for geography's existence, and that studying the geographical environment in an integrated way is the fundamental approach to dialectically understanding its formation and development. That is to say, only by adhering to an integrated direction can geography develop in the right direction.

As a discipline oriented toward practical application, the integrative nature of geography is reflected in its multidimensional and dynamic perspectives. Integrated geographical research, with the relationship between humans and the environment as its main thread, includes three main aspects: firstly, dynamic research focusing on surface and global environmental changes, that is, environmental dynamics research; secondly, human social dynamics research centered on human social development, with a focus on the relationship between environmental and social dynamics; thirdly, comprehensive analysis of research areas such as regions and basins. For example, in urbanization research, it is necessary to clarify the interactions and interdependencies between central cities and satellite cities in terms of industrial layout, transportation networks, and population distribution. At the same time, integrated geographical research not only needs to develop the theories and methods of integrated geography, but also needs to pay attention to the interdependence between scales. Geographical research spans different spatial scales, from the scale of ecosystems to the global scale. Studying regional characteristics and interregional dependencies is a major theme in integrated geographical research.

Integrated geographical research plays an important role in decision-making related to local, regional, and global environmental change and economic development. In response to numerous challenges concerning resources, the environment, ecology, and sustainable development amid China's rapid development, the advancement of Chinese geography needs to focus on major national demands, provide scientific support for government decision-making, achieve sustained development through application, and gradually form a distinctively Chinese school of geography. Currently, there are many examples of the application of geography in national decision-making, such as new-type urbanization, ecosystem protection and restoration, and watershed water resource management.

I. New-type urbanization

New-type urbanization requires consideration of the social elements of urbanization, the capacity of land and water resources to support urbanization, and the environmental carrying capacity. First, the core of urbanization is the urbanization of people. Urbanization research first needs to analyze social elements and identify how people gather in cities—without the clustering of people, urbanization cannot occur. Research on the social elements of urbanization includes how the population gathers in urban areas, the forces attracting this aggregation, which relate to the economic structure, employment structure, and industrial structure, as well as the spatial layout of urban settlements—in other words, the social processes of urbanization. Second, the development of urbanization should interact with natural resources, examining whether land and water resources can support urbanization. For resource-based cities, particular attention should be paid to the exploitation potential of mineral resources and the city's future transformation paths. In the relationship between urbanization and the environment, issues such as prevailing wind directions and sources of air pollutants need to be considered. Towns near water sources or lakes should also consider the degree of water pollution, soil contamination, and ecological degradation. Furthermore, human geography also studies spatial differentiation and development models of urbanization, such as the urbanization modes along the eastern coast, the fragile ecological zones of the west, and the rise of central China.

II. Ecosystem protection and restoration

Climate change and human activities are profoundly altering the structure and functions of ecosystems, while the mechanisms behind ecosystem processes and services still require deeper understanding. At small scales, mechanisms such as erosion control by forests or how runoff is generated within forest ecosystems can be studied through site-based monitoring. At the regional scale, however, beyond forests, other landscape types—farmlands, grasslands, urban areas—require integrated research methods combining surveys and remote sensing to analyze changes in spatial and temporal patterns, assess ecosystem services comprehensively, forecast future dynamics, and optimize spatial layouts. In studies on ecosystem protection and restoration, it is also necessary to pay attention to the coordinated development of ecological protection and regional economy, so that ecosystem services can sustainably provide benefits to human. When applying scientific research findings to government decision-making, it is necessary to integrate the mechanisms linking ecosystem structure, processes, and services into decision content, synthesize multi-scale research outcomes, and incorporate them into operable governance scales (e.g., city, county, or township), thereby serving urban–rural integration and regional sustainable development.

III. Watershed water resource management

Watershed water resource management often requires coordination of water allocation across the upstream, midstream, and downstream sections. The upstream is usually focused on ecological protection and water conservation; the midstream on agricultural development; and the downstream often on urban construction. The ecological and economic activities in the midstream and downstream, as well as ecological restoration in the upstream, all demand water resources. To achieve integrated allocation and protection of water resources within a watershed, it is necessary to study not only natural processes—such as ecological processes, surface water processes, groundwater processes, and cryospheric processes—but also socioeconomic processes, such as the market value of water and water resource allocation. The interactions and mutual constraints between natural and social processes, and how to balance and coordinate them, should also be addressed. Integrated watershed water resource management often requires the establishment of a decision support system to predict future changes. Such models typically involve hydrological, ecological, and economic components. For instance, the Heihe River Integrated Eco-Hydrological Research Program aims to couple hydrological, ecological, and economic models to build a decision support system for simulating and forecasting future changes. Watershed research need to proceed on multiple fronts. In terms of data, basic datasets are needed on vegetation, land use, climate, and socioeconomics, and should be integrated within study units. In terms of modeling, it is necessary to couple elements within the models, such as the relationships between water and the economy, water and ecology, and various hydrological processes. Based on this, a decision support system can be built to simulate, evaluate, and allocate water resources across different industries, ecological processes, and spatiotemporal scales.

Chinese geography is facing unprecedented opportunities. In the process of development, geographical research needs to deepen its comprehensive understanding of complex coupled human and natural systems through an integrative lens. Chinese scientists need to strengthen leadership and participation in major international research programs and enhance the study of global issues to advance the development of geographical science. To transition from a major country in geography to a global powerhouse, Chinese geographers should comprehensively improve the internationalization of Chinese geography. In August 2016, the Geographical Society of China successfully hosted the 33rd International Geographical Congress. As a quadrennial event, this was the first time in the 100-year history of the International Geographical Union that the Congress was held in China, and it was also the largest in its history, reflecting the growing influence and status of Chinese geography on the world stage. However, there remains a significant gap between Chinese geographical research and that of developed countries, and its level of internationalization still needs improvement. Future research needs to overcome theoretical and technical challenges, enhance the ability to interpret complex geographical phenomena, and more effectively realize the value of Chinese geography in serving scientific decision-making.

The development of Chinese geography must be deeply rooted in the major needs of the country. No other country in the world grants geography the academic status it holds in China. On the one hand, China's vast territory and uniquely complex geographical elements provide excellent subjects and testing grounds for geographical research. On the other hand, China is the most populous and one of the fastest-growing countries in the world.

The many resource and environmental problems brought about by the population explosion, economic development, and globalization present both unprecedented challenges and opportunities for the development of Chinese geography. Therefore, Chinese geographers need to not only strengthen frontier research in geography and innovate in developing comprehensive theories, methods, and technical approaches, but also focus on major national needs, provide scientific support for sustainable socioeconomic development, serve the coordinated and sustainable development of society, economy, and the environment, and gradually form a distinctively Chinese and internationally influential system of geographical science.

Originally published in Chinese in *Acta Geographica Sinica*, Issue 11, 2017, abridged and revised.

Geography for Global Sustainable Development

In 1962, the first major book focusing on environmental issues, *Silent Spring*, was published, awakening public awareness of environmental protection. In 1972, the first United Nations Conference on the Human Environment was held, adopting the "Declaration on the Human Environment". Since then, governments worldwide have increasingly prioritized environmental issues. Entering the twenty-first century, with the acceleration of globalization, environmental problems have become more pronounced. In 2000, the United Nations Millennium Summit established the Millennium Development Goals. In 2015, the United Nations Sustainable Development Summit released the "Transforming Our World: The 2030 Agenda for Sustainable Development", steering global development towards a sustainable path.

In the 1980s, international research on global environmental change began. In 2014, the International Council for Science and the International Social Science Council launched the 10-year "Future Earth" research initiative, aiming to guide a path to sustainable development.

Geography, from its inception, has been an integrated discipline. It studies the spatial patterns, temporal evolution, and regional characteristics of geographic elements (water, soil, climate, biota, and humans) and their integrated systems. It also explores the mechanisms of human–natural interactions near the Earth's surface. With characteristics of integration, interdisciplinarity, and regionality, geography seeks to "explore natural laws and reveal the humanistic essence". It offers multidimensional perspectives and integrated concepts, giving it a natural edge in advancing global sustainable development goals. Geography aims to address complex resource, environmental, and developmental challenges, focusing not only on interpreting the past but more importantly on serving the present and predicting the future.

At both regional and global scales, issues related to the environment, resource utilization, and sustainable development are becoming major challenges for human society. In recent decades, geography has rapidly advanced in theory, methods, and technology. Research paradigms have shifted from descriptive knowledge to the coupling of patterns and processes, and the modeling of complex human and earth system. Research methods have evolved from being primarily descriptive to becoming integrated and quantitative, transitioning from conceptual models to quantitative expressions, and from statistical models to simulation models. The research themes have moved from "pluralism" to "systems", with a focus on the Earth's surface system, analyzing and understanding the major issues society faces today. In terms of research approaches and methods, there is a focus on coupling geographic patterns and processes. The development of geography is increasingly playing an important role in providing a solid scientific foundation and technical support for global sustainable development, contributing to national economic, social, and ecological civilization construction.

The forefront and hotspots of geographical research focus on theories and methods of coupled human and natural systems, which require the development of element integration, process coupling, data fusion, and model development.

In terms of element integration, it is important to focus on the interactions of elements within coupled human and earth systems, early warning of the system's carrying capacity, and global integrated risk responses. Research will examine interactions between surface layers, human and environmental elements, the spatial characteristics of element associations, and the long-term evolution of these processes. Through various evaluation methods, the limits and tipping points of Earth's support capacity should be clarified. Early warning of global environmental changes should be achieved through system integration, with an understanding of the interconnections between risks, focusing on cascade effects, and filling knowledge gaps through interdisciplinary approaches. A multi-departmental response to networked risks is essential.

In terms of process coupling, attention should be paid to the comprehensive impact of human activities on climate change; sustainable use of food, energy, and water; biodiversity conservation and ecosystem services management; regional and global environmental pollution's health effects; sustainable urbanization and rural revitalization. Research will explore the impact of atmospheric composition changes, land surface properties, and anthropogenic heat releases on climate change. It will also evaluate and forecast human demands for food, energy, and water, proposing sustainable land use configurations for food, water, and energy security. The cascading relationships of "biodiversity–ecosystem structure–ecosystem processes and functions–ecosystem services" should be clarified to provide decision-making support for ecosystem management, and aimed at improving human well-being. Research on the effects of air, water, and soil pollution on human health is needed. A comprehensive research system should be established, integrating fundamental research, technological modeling, and policy consulting, focusing on issues such as urban–rural land resource allocation, the coordination of industrial layouts, the flow of production factors, and fairness in urban–rural policies and systems.

In terms of data fusion and model development, attention should be given to methods for processing massive geographical data and model systems. A multi-element Earth surface observation system, surface process coupling model system, and decision-support system for sustainable development should be established. The development of technologies and methods for geographic data fusion and integration is essential.

Geography continues to evolve. Its development is shifting from pattern research to process research, from element research to system research, from theoretical to applied research, and from knowledge creation to decision-making. Geography has become a fundamental discipline for sustainable development and will play a pivotal role in global and regional environmental and developmental issues.

Originally published in Chinese in *Science and Technology Review*, Issue 2, 2018, abridged and revised.

The Fusion of Nature and Humanity

Geography, as an interdisciplinary field committed to "exploring natural laws and revealing the humanistic essence", has, since its inception, focused on studying phenomena at the interface between the natural and human sciences. Here, the term "humanity" broadly encompasses various social, economic, political, and cultural phenomena, in contrast to "nature". As one of the three major branches of geography, human geography plays a key rolc in bridging geography and the social sciences.

Modern human geography in China originated in the 1930s, primarily in the form of regional geography. During the first three decades after the founding of the People's Republic of China, geographic research and education in China mainly focused on physical geography and economic geography. In contrast, human geography remained relatively stagnant, while Western human geography experienced rapid development in the 1960s and 1970s, giving rise to schools such as behaviorism, humanism, radicalism, and quantitative geography. This led to a growing gap between China and developed countries in the field. Since the 1980s, under the advocacy and promotion of geographers such as Professor Li Xudan and Professor Wu Chuanjun, human geography in China has experienced a period of comprehensive revival and advancement. Echoing the tremendous achievements of China's reform and opening up over the past four decades, various branches—including urban geography, rural geography, cultural geography, transportation geography, tourism geography, industrial geography, agricultural geography, population geography, social geography, political geography, and historical geography—have made significant progress. Guided by the concept of the regional coupled human and natural system, these branches are increasingly intersecting and integrating, providing scientific support for national and regional sustainable development.

Undeniably, as the branches of geography continue to develop and deepen, the discipline has begun to show signs of hollowing out—losing its integrative foundation as a unified field of study. Whether it is the trend of human geography becoming increasingly aligned with social science paradigms, or the shift of physical geography toward ecological specialization, both may undermine the integrative nature that defines geography as a discipline. If physical geography gradually becomes part of the natural sciences, while human geography restricts itself to social science theories and methods, the hollowing out of geography will only be exacerbated. The Future

Earth research initiative, launched in 2014, is a paradigmatic example of geography's evolution in the new era. It aims to foster the close integration of science and policy-making, and to bridge the divide between the natural and social sciences, thereby facilitating a global transition toward sustainable development. At the same time, as a scientific response to the pressing challenges of global sustainable development, the International Council for Science and the International Social Science Council merged in 2018—a move that exemplifies the growing trend toward the convergence of the natural and social sciences. Geography is the only discipline that truly bridges both the natural and social sciences. Preventing its hollowing out is essential not only for the advancement of the discipline itself, but also to meet the urgent demands of our time. Developing theoretical frameworks and methodological approaches for integrated geographical research is a crucial task—both to safeguard geography's integrative character and to establish its core disciplinary foundations. This mission requires the concerted efforts of contemporary geographers. As Professor Huang Bingwei emphasized—Integration is the basis for the existence of geography; the path to integration exists, but it must be arduously pursued.

The millennia-old traditions of rural China have shaped a unique human–natural relationship rooted in its agrarian civilization. In the modern era, the accelerated process of industrialization and urbanization has shaped the monumental and globally observed transformation associated with China's reform and opening-up. Chinese human geography research has been closely aligned with the strategic needs of national sustainable development. This includes major initiatives such as new-type urbanization, targeted poverty alleviation, rural revitalization, and the Belt and Road Initiative, as well as regional strategies like the coordinated development of Beijing–Tianjin–Hebei, the Yangtze River Economic Belt, the Guangdong–Hong Kong–Macao Greater Bay Area, and ecological protection and high-quality development along the Yellow River Basin. This task-oriented model, in which national development agendas drive disciplinary advancement, has made Chinese human geography more structured, standardized, rational, and authoritative compared to its Western counterpart. In contrast, Western human geography in recent decades has evolved around diverse interests and value orientations, giving rise to numerous schools of thought and streams of empirical research. As Professor Wu Chuanjun wrote in the preface to the first edition of *Introduction to Human Geography*—The Western human geography community is like a garden of contending schools and flourishing ideas, dazzling both observers and geographers alike. While such diversity has enriched the research content of human geography and strengthened its interdisciplinary connections with the humanities and social sciences, it has also undermined the recognition of disciplinary commonalities within geography and made integrative synthesis more difficult to achieve. In contrast, Chinese human geographers focus on solving practical contradictions between resource–environment constraints and socioeconomic development through urban, rural, and regional planning and policy-making. They place particular emphasis on applying integrated theoretical approaches to address the challenges of regional sustainable development—an approach that, to a certain extent, helps avoid the hollowing out of human geography and even geography as a whole. In this sense, China provides fertile ground for the advancement of human geography and the flourishing of integrative geographical research. This noble mission deserves the commitment of aspiring scholars who seek to elevate the value of geography and contribute to building a beautiful China.

In China's disciplinary classification system, geography is categorized as a natural science. This categorization tends to result in geography students at Chinese universities having stronger backgrounds in mathematics and physics, but also in a relatively weaker exposure to the humanities and social sciences. As such, there is a strong need for a human geography textbook that is theoretically rigorous, rich in content, broad in vision, and accessible in cognitive difficulty—one that can expand the knowledge base of geography and related majors while cultivating students' integrative thinking skills. However, under the ongoing trend of disciplinary differentiation, a growing number of specialized monographs based on specific branches of geography have emerged, whereas high-quality comprehensive materials remain relatively rare. *Introduction to Human Geography* is one of the few exemplary works of its kind. This publication covers a broad spectrum of topics such as population migration, industrialization, urbanization, globalization, locality, geopolitics, and cultural values. It also integrates Western theories of human geography with research grounded in Chinese realities. From a pedagogical perspective, it presents key concepts through extensive domestic and international case studies, offering vivid and accessible explanations. I believe that studying this book can meaningfully enrich students' knowledge systems in human geography and provide a strong foundation for future scientific research and real-world applications.

China's vast geographical expanse and the exceptional complexity of its physical and human landscapes are unparalleled in the world. This expansive territory provides not only a rich object of study but also a natural laboratory for geographical science. China is also the most populous country in the world and one of the fastest developing economies. Rapid population growth, economic development, and globalization have brought about numerous problems related to resources, the environment, and sustainable development, providing unprecedented opportunities and challenges for the development of geography in China. In response to the major strategic needs of ecological civilization construction and high-quality development in the new era, it is necessary to uphold the systematic view of harmonious coexistence between humanity and nature, to put into practice the regional development concept that "lucid waters and lush mountains are invaluable assets", and to implement the global governance vision of jointly advancing the construction of a global ecological civilization. In this context, human geography, together with physical geography, is poised to offer critical scientific support for sustainable development policy and practice at global, national, and regional scales. It is my hope that, guided by this publication, more students and practitioners will gain a deeper understanding of the integrated knowledge system of human geography. May this inspire a stronger interest in geography, encourage a conscious linking of personal scholarship with national development, and contribute to shaping the disciplinary spirit of geography while advancing our shared aspiration for the Chinese Dream.

When I received the invitation from Professor Gu Chaolin to write a preface for *Introduction to Human Geography* (Second Edition), I felt uneasy and unsure. As a physical geographer, I have only a superficial understanding of human geography. However, as someone who has in recent years consistently advocated for strengthening integrative research in geography and developing integrated geography, I cherished this rare opportunity to learn. After reading *Introduction to Human Geography* (Second Edition), from space and location to interaction and spatial diffusion, from population and ethnicity to the state, and from industrialization, urbanization, and globalization to industry and development, culture and values, geopolitics and global governance, I became more convinced that geography has a common foundation and an integrative conceptual framework, and my confidence in developing integrated geography was further strengthened. The above reflects my reading experience and serves as the preface.

July 2022 in Beijing

Excerpt from *Introduction to Human Geography* (Second Edition) Preface (originally published in Chinese)

"The Chinese Dream" and "The Soul of Geography"

The "Chinese Dream" is to realize the great rejuvenation of the Chinese nation, which is the greatest dream of the Chinese people since modern times. Its specific manifestation is the prosperity of the country, the revival of the nation, and the happiness of the people.

Achieving the "Chinese Dream" depends on the development of the whole society, and innovation is the primary driving force for development. General Secretary Xi Jinping pointed out that innovation is the soul of a nation's progress and an important force for the development of human society. To adapt to and lead the new normal of China's economic development, the key lies in transforming development drivers through scientific and technological innovation. Scientific and technological progress will fundamentally change the ways humans produce and live, and will profoundly reshape people's thinking and worldview. Scientific and technological progress also accelerates the process of modernization and sustainable development.

Science and technology have played a decisive role in the advancement of human civilization, pushing humanity from primitive civilization toward ecological civilization. However, the conflict between the demands of traditional industrial civilization and the limits of natural resource availability and environmental carrying capacity has become increasingly pronounced. Therefore, there is an urgent call for major innovations and breakthroughs in science and technology to promote industrial transformation and achieve ecological civilization. Ecological civilization is an important component of the "Chinese Dream", and geography is a powerful tool in supporting the construction of ecological civilization.

Geography studies the laws governing the interactions among various geographic elements in nature and society. It examines the spatial differentiation, temporal evolution, and regional characteristics of geographical elements and integrated geographical systems. Geography studies the mechanisms of interaction between humans and the environment on the Earth's surface, and is characterized by integration, interdisciplinarity, and regionality. Integrated geographical research focuses on the coupling of geographical elements and human–natural relationships. Building models and simulation systems is one of the approaches and methods in integrated geographical research.

Many fields of science and technology with potential for major breakthroughs today—such as energy and resources, population health, information networks, advanced materials and manufacturing, agriculture, and the environment—are closely related to geography and ecological civilization. For example, in the field of energy and resources: energy conservation, the clean and efficient use of fossil fuels, the development of renewable and new energy sources, and the exploration of deep-Earth resources. Also, in the field of advanced materials and manufacturing: molecular design, intelligent manufacturing, and the research and application of new materials will trigger revolutionary breakthroughs and transformations in industry. According to the carrying capacity of resources and the environment, the national territory spatial development pattern should be optimized, and the construction of the urbanization system should be promoted.

The construction of ecological civilization involves the comprehensive integration of economic development modes with various geographical elements. Effectively reducing the pressure that economic activities place on resources and the environment, making determined efforts to resolve overcapacity, accelerating industrial transformation and upgrading, and vigorously developing a circular economy; improving the quality and level of the environment, resolutely tackling air, water, and soil pollution, earnestly protecting and restoring ecosystems, and actively addressing climate change; establishing a resource-conserving society and promoting changes in resource utilization methods; strictly implementing energy conservation, emission reduction, and consumption reduction; and enhancing the economical and intensive use of water resources, mineral resources, and land resources—these embody the true soul of geography.

"Exploring natural laws and revealing the humanistic essence". Geographical research needs to be closely integrated with sustainable development decision-making and ecological civilization construction. Realizing the "Chinese Dream" and demonstrating the "Soul of Geography" also represents a significant opportunity for the development of geography in China.

Originally published in Chinese in *Teaching Reference of Middle School Geography*, Issue 15, 2017

A Dialogue on Geography with Middle School Students in Beijing

On the morning of November 19, 2017, I was invited to give a lecture entitled "Geography: From Knowledge, Science, to Decision Making Support" to middle school students in Beijing at the Academic Hall of the Chinese Academy of Sciences. This article records the Q&A session after the lecture.

Question: Hello, Academician Fu! My question is: the 19th National Congress proposed that "Lucid waters and lush mountains are invaluable assets". How can we reconcile the contradiction between economic development and environmental protection?

Answer: The idea that "Lucid waters and lush mountains are invaluable assets" is part of the Two Mountains Theory proposed by General Secretary Xi Jinping during his visit to Zhejiang. This theory is both scientific and forward-looking. When we talk about "lucid waters and lush mountains", we are mainly referring to the natural geographical environment—what we commonly call ecosystems—which include elements like mountains, water, forests, grasslands, and lakes. These ecosystems provide services that are valuable to humanity. Back in 1997, American ecological economists published an article in the prestigious journal *Nature*, estimating that the global ecosystem services were worth $33 trillion, which is 1.8 times the global gross national product (GNP) of that year. So what kinds of services do ecosystems provide?

The first category is the ecosystem service products directly related to humans. For instance, farmland, as an ecosystem, provides us with food. At the same time, ecosystems also provide us with fresh water, timber, fuel, etc.—what we call provisioning services.

The second category is regulating services, which are even more valuable. The very existence of life on Earth depends on the regulation provided by ecosystems. Mars, to this day, has no known life—no vegetation, no animals, and no soil. The regulating services of Earth's ecosystems help to stabilize the climate. For example, through the atmosphere and vegetation, they regulate temperatures suitable for human survival. In cities, if there is a forest or wetland area, the temperature in summer around these ecosystems is 2–3 °C lower than in areas without them. Another example is how ecosystems regulate floods. What we call water retention can be understood as ecosystems storing part of the rainfall during heavy precipitation—in the soil, in dead plant matter, or as groundwater—and later releasing it as gentle streams when there's a shortage. In fact, ecosystems do not generate more water—the total amount of water on Earth is balanced. What ecosystems do is regulate the spatial and temporal distribution of water. This is called water retention. When we need water, ecosystems supply it; during floods, they help reduce peak flows. From this, we can also derive ecosystems' capacity to reduce pollution. Many types of domestic sewage are treated through wetlands, which reflects the value of regulating services.

The third category is cultural services. We admire beautiful scenery—mountains and waters that are visually appealing. These landscapes provide us with aesthetic enjoyment and serve functions such as tourism and education. In our country, many tourist areas now treat tourism as a leading industry, which also drives economic development. A sizable tourism industry is directly supported by the cultural services of ecosystems.

In addition to the three types of services directly related to humans—those we can enjoy, there is also a fourth type called supporting services. These include biogeochemical cycling, nutrient formation, soil formation, and primary productivity. Although supporting services may not directly impact humans, their scale and strength are closely tied to the other three categories. Ultimately, if we convert these various types of services into economic value, we get what General Secretary Xi described as "Lucid waters and lush mountains are invaluable assets". this concept is scientifically grounded and also represents a practical summary based on real-world experiences.

On the negative side, if we destroy these lucid waters and lush mountains, the cost to restore them will be even greater. Here's a number for you: in 2008, the World Bank estimated that China had the greatest GDP impact globally due to environmental pollution and ecological degradation. Currently, our country's GDP growth rate is just over 6%. If we neglect environmental protection, all this growth could be offset or even be insufficient. From this perspective, the environment clearly has economic value. As for how to achieve coordinated development, this is a new issue. Three "red lines"—the red line for protection of farmland and permanent basic cropland, the red line for ecological conservation, and the boundary line for urban development—are all enforced through spatial regulation to promote coordinated development. Farmland must be protected to ensure ecosystems can continue to provide products for humanity. As a large country, food security is of utmost importance for China. Therefore, we must not destroy farmland or excessively occupy high-quality farmland. Similarly, the ecological conservation red line ensures that fragile ecological zones are protected. These forms of spatial control help achieve both economic development and environmental protection. At the same time, we must consider the carrying capacity of resources and the environment during development, and clarify the potential for development.

Question: Hello, Professor Fu! Among the various categories of geography courses in university, which one requires higher academic knowledge? Are there any less popular subjects?

Answer: Physical geography is more suitable for students with strengths in the natural sciences. It involves more mathematical and scientific knowledge and is mainly applied in environmental protection, ecological conservation, and sustainable resource use. Therefore, the major is called Physical Geography and Resource Environment, and it includes courses in mathematics, physics, and chemistry. During their studies, students also need to learn geographic information systems (GIS), remote sensing methods, and comprehensive knowledge on the interactions among atmosphere, soil, water, and other natural elements. For example, when studying ecology and hydrology, we don't look at them in isolation but focus on their interactions.

In the past, hydrology primarily focused on physical processes—rainfall, runoff, and rivers—without considering the role of biological elements. But today, ecology and hydrology must intersect. After rainfall, there are plants on the ground—some are shrubs, some are herbaceous. How do they intercept rainfall? How does infiltration occur? What roles do transpiration and evaporation play in the water cycle? Current research is much more detailed and integrates individual elements. This is physical geography.

Human geography is closer to social sciences, so now the major is called Human Geography and Urban–Rural Planning. It is more closely linked with urban planning. In cities, we not only need to consider transportation and roads but also the natural environment. Students majoring in physical geography might still take courses in human geography, and vice versa—human geography students need to study physical geography and also participate in fieldwork. Some students who are interested in urban planning can apply to architecture or urban planning programs. There are several schools of thought in urban planning, one of which comes from a geographical perspective. This school focuses not only on the details within the city but also on the relationship between the city and its surrounding environment, or between cities. Some urban planning approaches only consider block-level planning—such as the relationship between buildings and transportation. But from a geographic perspective, we also need to consider the interaction between cities and ecosystems.

I also hold the view that cities should integrate ecology into their planning, rather than merely using it to decorate cities. In urban planning, people often finish planning neighborhoods, roads, and traffic first, and then do the landscape planning—deciding where to plant some trees or grass. This is unscientific. Many cities today lack complete ecosystems because their planning lacks geographic and ecological perspectives. So, human geography is now combined with urban–rural planning. I suggest that if you want to do well in urban planning, start with a human geography undergraduate degree, and then study urban planning in graduate school. That way, you'll have a solid macro-level foundation and can later master more detailed knowledge, resulting in a more comprehensive understanding.

Question: Academician Fu, may I ask whether issues related to environmental pollution also fall under the scope of geographical research?

Answer: These belong to the scope of geographical research, but in addition to geographical research in universities, there is also a major called Environmental Science and Engineering. This major often focuses on the study of pollutant treatment, water purification, atmospheric pollution transmission, etc., involving more laboratory work. Environmental science can be more detailed and microscopic than geography, but when geography and environmental science are combined, in addition to analyzing the above aspects in the laboratory, field observations are also required, and the scale of research is larger. Geography studies the human living environment and focuses on changes across time and space. For example, geography pays more attention to atmospheric changes and pollution processes, while environmental science focuses more on the effects after pollutants are emitted.

Question: Academician Fu, what are the future development directions of geography, and how does geography contribute to decision-making, particularly in relation to population and resource issues?

Answer: The knowledge covered in geography courses can indeed play a role in decision-making. Nowadays, decisions are often needed for complex, interdisciplinary issues. Population, urban and rural development, ecological protection, and environmental pollution—all of these require informed national policies. These elements form a system—"pull one string, and the whole system moves". When we focus on a specific problem, the approach is to expand it into a system-wide analysis, and then propose solutions accordingly. So, comprehensive thinking and analysis in geography are extremely important. Take water issues, for example. They involve not just industry, population, and ecological usage, but also require a comprehensive analysis of all aspects of water use. Only then can we determine how to solve the problem. This is the analytical approach and method that starts from individual elements and leads to a comprehensive solution.

Question: I heard you mention GIS, including remote sensing and cartography, I am quite curious, because it is related to the IT industry. What are the future development advantages and prospects of this technology in geography? Or has it already reached a saturation point?

Answer: GIS began to thrive in the 1980s. Remote sensing solved the problem of data for large-scale spatial analysis in geography. Their emergence greatly promoted the development of geography. In the early days, traditional geographical research relied on field investigations and emphasized regionality. How far people could walk was how far they could see. Now, remote sensing observes the Earth from space, offering a very broad perspective and providing a wealth of data in terms of both time and space. The processing and analysis of these data require the use of GIS. Early on, we moved from hand-drawn maps to computer mapping, but it's not just about map display—GIS has spatial analysis functions, such as shortest path analysis. For example, when building a reservoir, one needs to know how large the flooded area will be. After calculating the water volume, you input the topographic map into the computer system to generate a DEM—a digital elevation model—which shows the spatial distribution of elevation. Then you input the water volume, and the computer graphics system immediately shows which areas will be submerged and which areas need relocation. It becomes instantly clear. Similarly, after sea level rises, how much area will be submerged and which types of land will be affected can also be made clear through GIS spatial analysis.

Therefore, GIS integrates mathematics and computer science—it is a new technology, method, and science intersecting with geography. Universities also have this major. Personally, I think it has basically reached saturation across the country. Many schools offer this program, but some schools that had no geography department set up GIS programs. This means they lack the thinking of spatial analysis and the training in basic geographic courses. As a result, these students only master the technical aspect and cannot play the advantage of geography in transmitting knowledge to science in integrated departments, as we mentioned earlier. In the U.S., GIS programs are still quite popular, especially at the master's level. Because students there are not divided into majors at the undergraduate level, they learn GIS technology and methods in one and a half to two years during graduate school. Many of them go on to manage GIS at the city or county level after graduation. In the U.S., at least every county has a GIS office, managing all spatially distributed geographic information within that county, including industry, transportation, protected areas, etc. It is expected that China will reach this stage of development in the future.

Question: After listening to your lecture, our understanding of the definition and research areas of geography has expanded. Population research is also a branch of geography. Now that our country has opened up the two-child policy, data shows that we will experience a population cliff in the next 20–30 years. What changes will happen after that? Are there corresponding studies or findings in geography?

Answer: Population issues fall under the field of population geography, which is a branch of human geography. In terms of China's research on population issues, relatively speaking, there is less research within the geography field. We have mainly studied the interaction between population and resources, known as the carrying capacity of resources—how many people such resources can support. Once, Lester Brown proposed the question of "Who will feed China?" and wrote a book about it. Geographers from the original Comprehensive Investigation Committee of Natural Resources also studied China's food issue. The result of carrying capacity assessments showed that feeding 1.6 billion people in China is not a problem. Lester Brown's book was published in the 1990s—more than 20 years ago—and China's development hasn't run into problems since. Chinese geographers studied the land resource carrying capacity and published a book in response called *Who Feeds America*. In its development, the U.S. has actually been using global resources and influencing the global environment.

As for China's population issue, the implementation of the two-child policy was considered from the perspective of population development. From the point of view of the demographic dividend, I think it was the right move. Because China is gradually entering an aging society, young people, as a labor resource, are the reserve and potential driving force for national development. From this perspective, we need to open up the two-child policy. Regarding population development, as you mentioned, there might be a "cliff", but I think it's actually a smoother process. We used to have strict population control, but in rural areas there were already many cases of two or three children, so the population policy doesn't follow a sudden cliff-like pattern as we imagine. Currently, both the demographic and geographic research communities need to start from a systemic perspective, combining social and economic development to seek solutions.

Question: I'd also like to ask more about GIS. Based on previous answers, I roughly understand that GIS is a technology and method for managing and analyzing data. You mentioned that data management and analysis has evolved from maps to GIS technology, and then to three-dimensional observation. You previously conducted two studies—one on regional environment and biodiversity conservation, and another on land use structure in hilly and gully areas. Could you talk about how GIS was applied in these two studies?

Answer: This student's understanding of GIS is quite accurate—it is a technology and tool for data management and analysis. But computer science also does data analysis and management. That's why, in GIS programs, besides learning computer science, databases, and software design, we also need to master essential geographical knowledge. Therefore, we should understand GIS as analyzing and managing spatial data with geographic thinking—that would be a more comprehensive view. We need to study to gain knowledge of planning, population, and the natural environment. So, the GIS major must include courses in geography, computer science, and mathematics. Only through this integration can one develop a complete knowledge structure in GIS.

Regarding your question, my main research is on the human–natural interactions. For example, on the Loess Plateau, we use GIS to conduct research from small to large scales. The small scale includes our monitored data, and the large scale involves remote sensing data. How can we integrate these datasets? We move from plot-based observation to transect-based surveys and then use GIS spatial analysis to scale up these observations to the regional level—that's the technical advantage of GIS. We combine GIS data with models for analysis to answer questions like: what kind of land use structure can minimize soil erosion on the Loess Plateau? To do this, we input our observed results and related data into the GIS for analysis. For example, for farmland with slopes over 25° that needs to be revegetated—where should be planted with grass, and where should be afforested—this layout must be determined based on land suitability evaluation, which can be achieved through GIS.

Question: You mentioned that China is currently facing many environmental problems. I would like to know what kind of policies and measures are being taken domestically to solve these problems simultaneously. With so many problems piled up, is the situation now quite serious?

Answer: We are indeed facing these problems now, but the government has also taken a series of measures in the past five years to address them. For example, the division of main functional zones has been completed from the national level down to each county. From the National Development and Reform Commission to the provincial and county levels, this framework is being used. Main functional zones identify key development areas, where economic development is encouraged, while restricted development areas require nature conservation. The delineation of restricted zones is often combined with ecological red lines. In development, these red lines must not be crossed—areas that need protection should be protected, and those that need restoration should be restored. On the national scale, many areas in Xinjiang are designated as restricted development zones, but within Xinjiang, some parts of those areas may still be developable, which requires further refinement. In such regions, certain development is allowed. This is a national-level, macro-level, territorial control measure.

As for environmental pollution, even more policy measures have been implemented. The government has adopted many supervision and monitoring methods. Environmental protection staff have been dispatched to local regions to supervise and ensure that sewage and exhaust emissions are not discharged illegally. For air pollution control, cities like Beijing have implemented vehicle restrictions, and vehicle emissions must be monitored every six months. From a policy perspective, the Ministry of Environmental Protection issued the "Ten Measures for Air", "Ten Measures for Water", and "Ten Measures for Soil". These aim to control air, water, and soil pollution, and are strictly enforced in various regions. In recent years, some environmental conditions have already improved. Environmental pollution is also closely related to economic transition—some high-energy-consuming and heavily polluting enterprises must be gradually phased out, especially some small factories. Through technological and industrial innovation, new industries should be developed to replace them. Meanwhile, pollution treatment also needs to be carried out—not only controlling it at the source but also at the end. For example, flue gas emissions from boilers and vehicle exhaust need to be equipped with purification devices. Measures are being implemented through policy, technology, and government management at all levels. Although there are many problems, the current environmental protection measures are also quite powerful.

Question: You just mentioned that the focus of research in China lies in the economy, which differs somewhat from abroad. Comparing research in China with that in other countries, how does China integrate and make use of foreign research to address domestic issues? Does university geography involve these economic-related topics?

Answer: There are indeed some differences between human geography research in China and that in the West. We are currently in a phase of rapid economic growth, so Chinese human geographers tend to focus more on economic, urban, and tourism–related issues, such as urbanization. Since about 90% of the population in developed countries already live in cities, they don't pay as much attention to urbanization. Instead, they focus on issues like gender—for example, the role of women in industries—and social equity. However, we also learn from some of their research methods. As social equity issues have emerged in China, we have begun to study geographic topics related to society and culture. In the geography department of universities, as I just mentioned, there are three majors. Among them, the human geography major is closely related to economics. Within human geography, the area most closely tied to economics is regional economics. That's why we call the program "Human Geography and Urban–Rural Planning". To study how the economy and industries are arranged across spaces, cities, and regions, we must focus on the relationship between economy and geography.

Question: Concerning many modern human challenges—for instance, the population base is now very large, and population aging is severe. Because people live longer now, if we continue to have many children, the Earth will become overcrowded and humanity will face even more problems. If we are to ensure the sustainable development of the Earth, how can we maintain the population at a relatively balanced level? That is, keeping the population size stable while allowing people to live long lives. We cannot let it grow unchecked, nor can we reduce it too much, otherwise we'll encounter the problems I just mentioned.

Answer: This is once again a population issue. In fact, this is a systemic issue—population problems involve land use and the global environment. To determine whether the Earth can carry and support such a large population, we need to consider that people are creative beings, not just resource consumers. People also generate a lot of productivity. Therefore, we believe a large population can create a demographic dividend and stimulate economic growth, so population should reach a reasonable size. Just imagine: if our country only had 300 million people, life might be quite comfortable. But now think about it—if we didn't have over one billion people, our economy would not have grown this fast. So population is a double-edged sword: it can both create productivity and consume resources. Therefore, we need systemic research to both harness the potential and benefits of population and reduce its consumption of resources and impact on the environment.

In fact, with technological advances, the question of population carrying capacity is no longer merely calculated based on natural resources and the environment. Previously, evaluations of population carrying capacity had boundaries—for example, the carrying capacity of one specific country. Now we need to consider it at the global scale, using resources from the global market. Take Japan as an example: its usable land area is small, and its population density is not low, yet it enjoys a high *per capita* life expectancy and GDP. Therefore, we need to create a pattern and model of coordination and mutual advancement between humans and the environment, humans and society, and humans and the economy to promote human development. Population issues need to be viewed from both sides.

Question: Hello, Academician Fu. I am a student from the High School Affiliated to Beijing Normal University and I'm from Xinjiang. Will large-scale land reclamation in Xinjiang cause problems for the environment?

Answer: Xinjiang is indeed an ecologically sensitive and fragile region. Regarding the issue you just raised about large-scale land reclamation in Xinjiang—personally, I believe Xinjiang's land should be treated as a reserve arable land resource for our country. At this stage, it is not suitable for large-scale reclamation. We need to assess the water resource situation before reclaiming land. That's because Xinjiang is an arid region with limited water resources, and many rivers rely on glacial meltwater from high mountains. In the coming decades, as temperatures rise, many glaciers—like Glacier No. 1 at the headwaters of the Urumqi River in the Tianshan Mountains—retreat year by year. If temperatures continue to rise, the potential of river water resources is limited. So we need to reclaim land based on the available water resources, to avoid waste after development. But such land can still serve as a national reserve resource.

Question: On global climate issues, there was the Kyoto Protocol. When it expired, a global conference was held in Copenhagen, where the Copenhagen Accord was established. But the United States chose to withdraw. What impact does the U.S. withdrawal have on itself and on other countries?

Answer: What you said is very accurate. Let me add that the agreement was not actually reached in Copenhagen. It was two years ago in Paris when developed and developing countries finally came to an agreement. China participated in this agreement and committed to reducing energy emissions by a certain proportion. There is a distinction between responsibilities for developed and developing countries. Although all must take responsibility, their responsibilities are differentiated. This gives developing countries room to grow and acknowledges the long-term emissions of developed countries.

After the new U.S. administration came to power, it withdrew from the Paris Agreement. The U.S. withdrawal has had some small impacts on other countries, but not significant ones. European countries have consistently paid attention to climate change and advocate for reducing greenhouse gas emissions. So, within the West, there are different camps on climate change. The U.S. withdrawal won't affect the overall picture. China has taken on the responsibility of a major power and is the backbone of the Paris Climate Agreement. This situation presents both challenges and opportunities for China. Reducing emissions during economic development will somewhat affect the pace of growth, but it also provides opportunities for economic transformation. We need to combine climate issues with our current efforts to address environmental pollution. Reducing pollution is also a path toward solving climate problems. Preventing water and air pollution and lowering energy consumption will also promote the transformation of our industries. Now, we need to develop emerging industries like electronics, information technology, the internet, and biopharmaceuticals to replace traditional industries. The high-pollution industries of the past will gradually be phased out. This is also the opportunity that the climate agreement has brought us.

Host: Thank you very much, Academician Fu, for answering the students' questions with such care. This has been a presentation with a broad vision and high-level perspective that is also closely connected to our living environment. Since Academician Fu studies geography, he used geographic thinking to navigate a complex world from a high-level and broad perspective. Geographic thinking is integrated and systematic thinking. So today's insights from Academician Fu will have a profound impact on everyone.

What Academician Fu has presented to us is geography as a foundational and directional discipline that concerns the future fate of humanity. What kind of discipline is geography? On the surface, it seems to encompass all kinds of knowledge. It may be the most knowledge-rich discipline in the history of human development. But in essence, it is a discipline of thinking—a discipline that trains one's thinking system and builds cognitive structures. Geography can represent an iceberg, or it can represent an elephant.

We rarely ask ourselves: which part of the "elephant" does the knowledge we're learning pertain to? Without asking this, humanity may take many detours. But if there's a discipline that lets you understand what the whole "elephant" looks like, and then you study each part and feel it out, you'll have both a holistic understanding and a detailed analysis. Geography can help you recognize the full iceberg of the global human living environment and then study it piece by piece.

Geography is a training of comprehensive thinking because it studies complex relationships—between people and nature, and between people and society. If you can clarify these relationships, you're surely a smart person. Once you build the habit of systematic and integrated thinking, it can help you grasp the direction of the world's development.

Geography can also unite various disciplines and put their core principles and knowledge to work for the world. Geography is therefore also "grounded"—it allows you to discover complex problems. It's a discipline that can go as low as discovering local issues and as high as guiding the future direction. That's why this discipline is so important.

Academician Fu's presentation gave us great inspiration and brought us into a future world. In the past, because there were fewer people, humanity recklessly exploited the Earth's resources. Now, as the population grows and resources become relatively scarce, we have entered a new era. Previously, we needed to inventory what we had—now we need to organize and plan what we've inventoried. Whether we like it or not, this era has inevitably arrived. So we need clear minds to understand how humans and the environment can coexist in harmony. Academician Fu's presentation gave us a lot to digest, and we hope everyone will think deeply about what our future world and life should be like. Finally, let's give a warm round of applause to thank Academician Fu.

Originally published in Chinese in *Lectures by Renowned Scientists* (Volume 17), 2018

Space and Process

Spaces and places are descriptions of geographic differentiation and local characteristics, serving as the foundation of geography. While classic geographical works have systematically discussed space and place, but merely describing spatial differences and local characteristics is insufficient. Geography must delve into the study of processes. This includes focusing on the shaping and evolving processes of spatial differences and local characteristics, the spatial interactions between geographic elements, and the coupling of geographic patterns and processes.

Coupling of Patterns and Processes

Since its inception, geography has been studying phenomena at the interface between the natural sciences and the humanities, making it an interdisciplinary discipline. Geography is distinguished by its integrative nature and regional characteristics. Its integrative nature is reflected in the diversity of elements it encompasses, while its regionality manifests as spatial differentiation or regional differences. However, with the development and deepening of sub-disciplines, geography has exhibited signs of becoming hollowed out. Professor Ron Cooke, a former president of the Royal Geographical Society, pointed out: "Geographers should strengthen basic research and establish a core theoretical foundation for geography. In interdisciplinary research, geographers need to hold their ground, continue contributing to society through applied studies, and strengthening basic research is the only way to achieve this."

Developing theories and methods for integrated geographical research is a crucial task in preventing the hollowing out of geography and in building core geographical theory. However, the greatest challenge in geography is also integration. As early as the 1960s, senior geographers such as Professor Huang Bingwei emphasized that integration is the foundation of geography's survival, but superficial integration

B. Fu, *The Soul of Geography*, https://doi.org/10.1007/978-981-95-1062-7_2

is untenable and must be based on in-depth disciplinary studies. Professor Huang proposed conducting research on physical, chemical, and biological processes within the geographical environment and integrating them, thus establishing three important frontiers in physical geography: the surface water–heat balance, chemical element migration, and biogeographical communities. These directions pioneered theoretical and experimental studies on physical geographical processes in China. Such ideas have had a profound influence on deepening process research and strengthening mechanistic exploration in Chinese physical geography. Both natural environments and human activities exhibit spatial heterogeneity. Studying regional geographical processes and their effects under such heterogeneous conditions represents a research frontier in geography.

The regionality of geography is expressed through spatial differentiation, or "patterns", while "geographical processes" show the spatiotemporal evolution of geographical phenomena. Coupling of patterns and processes is an important approach and method in integrated geographical research, as well as a breakthrough point for studying surface processes. Patterns and processes typically refer to the spatial relationships and evolving processes of different geographic or landscape units. In terms of patterns, they can be described by size, shape, number, type, and spatial configuration. These descriptors of pattern have inherent geographical significance. For example, patches of different sizes provide different ecological and resource domains, which are critically important for biodiversity conservation. Likewise, patch shape can influence the movement of water, soil, and organisms, and the number of patches can indicate the degree of landscape fragmentation. Furthermore, describing patterns from the perspective of spatial configuration can reflect their spatial structure characteristics, zonal and non-zonal laws. In terms of processes, they can be divided into natural processes (e.g., the distribution and migration of elements and water, species distribution and migration, runoff and erosion, energy exchange and transformation) and socio–cultural processes (e.g., transportation, population dynamics, and cultural diffusion). Patterns serve as the carriers of processes, while ecological processes encompass numerous driving forces that shape patterns. The two are inseparable, and their relationship can be expressed as "patterns influence processes, and processes change patterns". In specific research issues, it is often necessary to couple the two. For instance, differences in landscape structure or land use patterns can affect soil moisture and nutrient cycling, leading to a series of ecological and environmental effects such as land degradation and soil erosion. Conversely, hydrological processes like floods and surface runoff can rapidly change surface landscape patterns. Therefore, scientific research needs to couple the two—neglecting either side will result in an incomplete understanding of the research object. Depending on the scale of the study, the coupling of patterns and processes is mainly achieved through two approaches: direct observation and systematic analysis and simulation. Direct observation is typically conducted at smaller spatial scales and relies on specific observational and experimental techniques to accurately capture the relationship between patterns and processes. For example, to study differences in water retention among different vegetation types, real-time monitoring of soil moisture and temperature under typical vegetation types needs to be conducted at the plot or slope scale. At larger spatial scales, the coupling of patterns and processes often involves multiple natural elements and processes and is inherently complex. In such cases, systematic analysis and simulation methods can be employed. At present, the coupling of patterns and processes has become an important pathway for studying the mechanisms of natural processes such as carbon–water cycles, ecohydrological processes, and soil–vegetation–atmosphere interactions.

Originally published in Chinese in *Acta Geographica Sinica*, Issue 8, 2014, abridged and revised.

Land Use Patterns and Ecological Processes

"I often dream of returning to Yan'an, with both hands embracing the Mountain Baota." The bold and passionate rhythm of the folk tune *Xintianyou* is a beloved melody in the hearts of all who live on the Loess Plateau. Just a few dozen kilometers from Mount Baota lies Yangjuangou, located in the typical hilly and gully area, making it an ideal site for studying land use patterns and ecological processes. As early as 1996, after on-site investigation, I selected this area as a research base. Leading a team of researchers and students, we remained committed for over a decade, conducting systematic research on land use patterns and ecological processes at the small catchment scale.

Life in Yangjuangou was somewhat tough. The students spent more than half the year there, living in cave dwellings rented from local villagers, with five or six people squeezing onto a single *kang* bed-stove. The instruments and equipment were kept in another room. To make a phone call, one had to hike up the mountain to find signal. To take a shower, it was necessary to go all the way into town. Water and power outages were a common occurrence. However, whenever it rained, the students would get excited. While the villagers rushed downhill to avoid the storm, the students had to climb uphill. To collect data on runoff, sediment, stemflow, and precipitation redistribution after rainfall, they became "storm chasers", eagerly awaiting rain. Based on long-term observations, these "storm chasers" gained valuable insights through their perseverance and wisdom.

Land Use Patterns and Soil Moisture

The Loess Plateau lies in arid and semi-arid regions, where soil moisture is the main limiting factor for vegetation growth. Soil moisture is influenced not only by natural processes such as precipitation and evapotranspiration, but also closely related to land use types. Observations in the Yangjuangou catchment showed significant differences in soil moisture under different land use types: forest land had the lowest soil moisture, followed by grassland, with sloping cropland having the highest. Moreover, the spatiotemporal dynamics of soil moisture at different soil depths were closely related to land use types. In dry years, differences in soil moisture between land use types were more pronounced. In water-scarce areas, fast-growing plantations may cause soil desiccation within the first 20 years. Taking into account key factors such as slope, land use, vegetation cover, and surface runoff characteristics, we developed a model for the spatial distribution of soil moisture suitable for the loess hilly slopes, and validated it using field data.

Land Use Patterns and Soil Nutrients

Soil nutrients are a fundamental basis for vegetation and crop growth, and also key indicators of land degradation and productivity. Different land use types and their spatial arrangements affect the distribution and transport of soil nutrients like nitrogen and phosphorus. Our observations and analyses of soil organic matter, total nitrogen, available nitrogen, and available phosphorus under different land use types on the Loess Plateau revealed that forest land, shrubland, and grassland had higher nutrient levels than fallow and cultivated land. Studies on the spatial and temporal variation of soil nutrients under various land use patterns showed that converting sloping cropland to natural grassland or shrubland significantly improved the capacity for soil organic carbon sequestration. Compared to sloping cropland, the organic carbon density in afforested land, grassland, and orchards increased by 29%, 28%, and 28%, respectively.

Land Use Patterns and Soil Erosion

Different land use types can influence soil erosion by altering the underlying surface characteristics. Our research found that, under the same rainfall and slope conditions, sloping cropland suffered more severe soil and water loss than pasture, arbor forest, or shrubland. Long-term observations in the Yangjuangou catchment showed that due to the reduction of cropland and the expansion of forest land, average soil erosion in the 1990s was 24% lower than in the 1980s. However, the impact of land use change varied over space and time. For example, in the early stages of farmland abandonment, vegetation recovery was accompanied by soil compaction, leading to increased surface runoff and no significant reduction in erosion within runoff plots. Over time, the cumulative runoff reduction in shrubland and natural grassland gradually increased. Moreover, different land use combinations had significant impacts on soil erosion. Research showed that land use combinations such as "grassland + mature forest + grassland" and "grassland + young forest + mature forest + grassland" were more effective in controlling soil erosion; compared to the "grassland + shrubland" combination, soil erosion was reduced by 42%.

Rational Allocation of Land Use Structures in the Loess Hilly and Gully Region

Based on an in-depth understanding of the relationships between land use patterns and soil moisture, nutrients, and erosion processes, we explored rational allocation of land use structure in loess hilly slopes. In the Yangjuangou catchment, we compared representative land use spatial combinations for their ability to retain soil moisture and nutrients. From the hilltop to the bottom, the land use combinations were: "forest–cropland–grassland", "forest–grassland–cropland", and "grassland–forest–cropland" (see figure below). Field sampling and analysis showed that the "forest–grassland–cropland" combination had the strongest capacity to retain soil nutrients, and had lower soil moisture in the early stages, which was the optimal land use structure for soil and water conservation on the loess hilly slopes.

Persistence and Adherence

Our research aims to understand natural laws and the mechanisms of interaction between human activities and nature. The research process is long, and persistence and adherence are essential. Our work is conducted primarily in the field, not in laboratories or simulated environments. Each year, students go out to monitor rainfall, runoff, and other variables. Unexpected results sometimes occur during field monitoring. In the early 2000s, the Loess Plateau experienced two consecutive years of relatively low rainfall, with no runoff events, making it impossible to obtain relevant data. One student, after several years of monitoring, had only a few data points. When submitting his paper, reviewers commented that the dataset was too limited. I encouraged him to keep going—only through long-term persistence can good results be achieved. The student persisted for another year, which happened to be a wet year. After obtaining a large amount of data, he compared the wet and dry years, which yielded some good results. Not only did he supplement the original paper with new data, but also analyzed the differences between dry and wet years. The paper was soon accepted by an international journal.

Land use patterns and changes on the slopes in Yangjuangou

Ecosystem Service Trade-offs and Synergies

Ecosystem services are the foundation of human survival and development. They not only provide humans with abundant products but also support and maintain the Earth's life-support system. Ecosystem services depend on the structure and processes of ecosystems. A deep understanding of the formation and interaction mechanisms of ecosystem services is a key scientific requirement for improving ecosystem management and promoting regional sustainable development. The Loess Plateau is a typical ecologically fragile region and a climate-sensitive area. It is an important water source area of the Yellow River and the primary source of most of the river's sediment. A rational layout of ecological construction on the Loess Plateau that coordinates ecosystem water consumption with soil conservation and carbon sequestration is of great significance for the sustainable development of both the Loess Plateau and the Yellow River Basin.

Building upon research into land use patterns and ecological processes, we closely integrated the scientific frontiers of ecosystem services with the national needs of ecological restoration on the Loess Plateau. Through a combination of field observation and investigation, remote sensing monitoring, and model development, we conducted integrated research that linked ecosystem processes and services, and clarified the mechanisms of trade-offs and synergies among ecosystem services.

Our research found that the trade-offs or synergies among ecosystem services on the Loess Plateau exhibit significant nonlinear characteristics and vary with different types of vegetation restoration and the duration of restoration. With vegetation restoration, ecosystem carbon sequestration and water yield tend to exhibit a trade-off relationship, while carbon sequestration shows a synergistic relationship with hydrological regulation and soil conservation. Water is the primary limiting factor for ecological restoration in arid and semi-arid regions. While vegetation helps conserve soil and reduce sediment flowing into the Yellow River, the trade-off between carbon sequestration and water yield poses significant challenges to the sustainability of the Grain for Green Program. Our multi-scale research quantitatively revealed the trade-off relationship between carbon sequestration and water yield during vegetation restoration, along with their spatiotemporal variations. Observations at the plot scale showed that vegetation restoration can effectively enhance ecosystem carbon sequestration services. However, unreasonable ecological construction layout has also led to severe water deficits and the formation of soil desiccation in some areas, which in turn negatively affects vegetation growth. Research based on remote sensing and model simulation reached similar conclusions. At the regional scale, both net primary productivity (NPP) and net ecosystem productivity (NEP) on the Loess Plateau increased steadily. From 2000 to 2008, the average carbon sequestration rate reached 19.2 g C m^{-2}, with a total of 96.1 Tg C sequestered, transforming the Loess Plateau ecosystem from a net carbon source in 2000 into a net carbon sink by 2008. Quantitative research on the relationships among carbon sequestration, soil moisture, and runoff following vegetation restoration showed that, from 2000 to 2010, the Grain for Green Program led to reductions in soil moisture (2.4 ± 0.9 mm a^{-1}, $P = 0.08$) and runoff (0.5 ± 0.3 mm a^{-1}, $P = 0.006$) across the Loess Plateau. These reductions were equivalent to $67 \pm 53\%$ of the increase in water consumed by carbon sequestration (i.e., evapotranspiration).

Due to the trade-off between carbon sequestration and water yield services, quantifying the vegetation carrying capacity of the Loess Plateau is essential for implementing ecological restoration appropriately in terms of timing, location, and intensity. By constructing an integrated framework coupling the natural–social–economic dimensions of water resource utilization on the Loess Plateau—and considering both vegetation water consumption and socioeconomic water use—we calculated the regional vegetation carrying capacity under different scenarios based on the carbon–water trade-off relationship. Results show that the maximum vegetation carrying capacity threshold—i.e., the level of vegetation productivity that could be supported if all annual precipitation is used for vegetation growth—is 445 ± 105 g C m^{-2} a^{-1}. The sustainable regional vegetation carrying capacity—i.e., the productivity level achievable after reserving water for local socioeconomic use—is 395–405 g C m^{-2} a^{-1}. Currently, vegetation productivity on the Loess Plateau is about 400 g C m^{-2} a^{-1}, which is already close to this threshold. Timely and site-appropriate afforestation is key to regulating the relationships among ecosystem services such as soil conservation, carbon sequestration, and water yield in the region, while grass–shrub vegetation or natural regeneration is more conducive to the sustainable management of ecosystems.

Constructing a regional integrated ecosystem services model based on the quantitative analysis of various ecosystem services can reveal the impact of human activities on ecosystem services and enable scientific ecosystem management. For this purpose, we independently developed a spatially explicit assessment and optimization tool for regional ecosystem services (SAORES), tailored to the context of the Loess Plateau. This model features capabilities for evaluating the mass and value of ecosystem services, simulating land use/land cover and management scenarios, and optimizing decision-making. It includes a scenario development module, an integrated ecosystem service model base, an ecosystem service trade-off analysis module, and a multi-objective spatial optimization module based on NSGA–II (Non-dominated Sorting Genetic Algorithm II). Based on the calculated results of ecosystem services, the model applies multi-objective genetic algorithm optimization to propose regional ecosystem management plans. Using this model, we assessed major ecosystem services—such as water retention, soil conservation, carbon sequestration, and food production—under different Grain for Green scenarios in a typical watershed of the Loess Plateau. We obtained land use configurations and regional ecosystem management strategies that maximize key ecosystem services, thereby optimizing the Grain for Green strategy.

Related results were published in *Nature Climate Change* and other journals.

Where Has the Sediment Gone?

The Yellow River is the river with the highest sediment load in the world. Over 90% of its sediment comes from the Loess Plateau in the middle reaches. In the past, the Loess Plateau was described as "bare-headed hills and muddy gullies"—a heavy rain would wash away soil, and planting a crop would strip off another layer of earth. After decades of ecological restoration and management, the landscape is now lush and vibrant. The annual sediment load of the Yellow River has decreased from 1.6 billion tons in the 1970s to about 200 million tons today.

The once "endless sediment" of the Yellow River now seem to be turning into a "clear Yellow River". So, what has caused the river to become clearer? Where has all that sediment gone?

Is it climate change or human activity? Which factor plays the dominant role? If it's human activity, what are the respective contributions of vegetation-based and engineering-based measures?

Our analysis shows that changes in sediment are closely related to precipitation, water yield capacity (the ratio of river discharge to precipitation) and sediment concentration (the ratio of annual sediment mass flux to discharge). Based on this, we developed an attribution approach to quantify the contribution and effect of each factor. Using 60 years of precipitation, runoff, and sediment load data from the Loess Plateau stretch of the Yellow River, we found that 58% of the reduction in sediment load was due to the decline in water yield capacity, followed by the contributions of sediment concentration and precipitation, accounting for 30% and 12%, respectively.

In more specific time periods, landscape engineering, terracing and the construction of check dams and reservoirs were the primary factors driving 54% reduction in sediment load from the 1970s to 1990s. Since 2000, with the implementation of the Grain for Green Program, large-scale vegetation restoration has become the main contributor, accounting for 57%. Therefore, the main reason for the reduction in sediment in the Yellow River in recent decades is human activity. Maintaining a sustainable ecosystem on the Loess Plateau plays a critical role in the stability of Yellow River's water and sediment. Future water and sediment management of the Yellow River needs to shift from comprehensive management of small watersheds in the Loess Plateau to coordinated management across the entire basin.

This research was published in *Nature Geoscience* in 2016.

Spatializing Global Nature's Contributions to People

The characteristics of the natural environment and human activities in geographical research both exhibit spatial heterogeneity, which necessitates the use of geographical regionalization systems and zoning methods to identify the spatial regularities of similarity and difference among regions. In the 1960s and thereafter, the older generation of geographers in China conducted systematic natural geographical zoning of the country, which had significant impacts both domestically and internationally. With global and regional economic development and the intensification of human activities, natural ecosystems are being increasingly disturbed by human interventions, and ecological and environmental problems at both global and regional scales are worsening. There is an urgent need to carry out ecological zoning research. Ecological zoning originates from geographical zoning and refers to the spatial division or integration of ecological units. For example, in 1989, the American scholar Professor Robert G. Bailey proposed the first globally comprehensive terrestrial ecological zoning scheme. In 2001, our team proposed the first ecological zoning scheme for China, based on the characteristics and functions of ecosystems, dividing China into 3 ecological macro-regions, 13 ecological regions, and 57 ecological zones.

Although global ecological zoning schemes have continued to emerge, most of them are based on biogeographical classification. The ecological zones delineated in these schemes are generally regarded as spatial representations of biodiversity, but they fail to reflect the multifunctionality of ecosystems and the ecosystem services that humans derive from them. This calls for the establishment of a technical framework for ecological functional zoning and the development of the world's first ecological functional zoning scheme, which can be applied to the sustainable development and integrated management of ecological resources at both global and regional scales.

Through extensive research and in-depth comparison, we found that the most advanced and widely recognized indicator system for assessing ecosystem functions in the international scientific community is the set of 18 indicators of Nature's Contributions to People (NCP), proposed by the Intergovernmental Science-Policy Platform on Biodiversity and Ecosystem Services (IPBES) established under the auspices of the United Nations. However, a major challenge lies in the fact that both the *Global Assessment Report on Biodiversity and Ecosystem Services*, released by IPBES in 2019, and scientific articles in top international journals—including

Science, *Nature*, and *Proceedings of the National Academy of Sciences* (*PNAS*)—have only conducted spatial mapping for a limited number of NCP indicators. As a result, it remains impossible to gain a comprehensive understanding of how much nature contributes to people at each location across the globe, where the high-value areas of nature's contributions are, and whether different types of NCP indicators exhibit synergies or trade-offs.

Based on the conceptual framework of NCP proposed by IPBES, we propose a technical framework that links biogeographical zoning with ecological functional zoning, construct an indicator-based evaluation method, and integrate multi-source spatial datasets on ecosystem evolution, climate change, and human activities at the global scale. For the first time, we conduct global spatial mapping of all 18 NCP indicators from 1992 to 2018, and identify the dominant combinations of NCPs within 15,204 watersheds. We delineate and define ecological functional zones based on two types of NCP indicators—high-value regulating and material NCPs, and identify the trade-offs and synergies among NCPs under global landscape change.

The results show that high-value areas of most NCP indicators appear only at a local scale across the globe. From 1992 to 2018, 6 regulating NCPs, 3 material NCPs, and 2 non-material NCPs declined. These reductions in NCPs can be largely attributed to the loss of natural ecosystems. In contrast, the increases in other global NCPs are mostly not the result of ecosystem recovery, but are instead driven by global warming and increased intensity of human resource use. Synergistic changes in NCP, characterized by co-increase or co-decrease and driven by landscape pattern changes, are more common than trade-off relationships characterized by opposing increases and decreases. However, some ecological functional zones located in transitional climate regions not only exhibit a limited range of NCPs, but also show strong trade-off relationships.

The global ecological functional zoning technical process we proposed is the first international example of ecological functional zoning developed based on the conceptual framework of NCP. The research findings have received close attention from Professor Bailey. Our assessment represents the first global study, following the release of the IPBES report, that maps all 18 NCP indicators and analyzes their trade-off and synergy relationships. It has significantly advanced understanding of the global spatial heterogeneity of this indicator system.

Related findings were published in 2023 in *Science Bulletin* and other journals.

Systems Thinking

The core of geographical research is the coupled human and natural system, in which human and natural components interact. It is characterized by complexity, non-linearity, and uncertainty. Geographical research requires systems thinking and an integrated approach to deepen the understanding of complex coupled human and natural systems, reveal dynamic evolution of socio–ecological element, analyze coupling mechanisms among processes, clarify optimal regulation pathways, and serve regional sustainable development.

How Does Quantitative Change Lead to Qualitative Change?

"Desert smoke rises straight, the long river sets round." This line of poetry by Tang Dynasty poet Wang Wei depicts the unique and magnificent scenery of China's northwest frontier. The arid and semi-arid regions of Northwest China are vast, with scarce precipitation, high evaporation rates, sparse vegetation cover, and poor soil quality, rendering its environment exceptionally fragile and sensitive to both climate change and human activities. When the external environmental pressure on dryland ecosystems exceeds a critical limit, it can abruptly shift from a stable state to a degraded one, a phenomenon known as crossing the "ecological threshold". Increasing aridity and grazing pressure are usually identified as primary drivers of these abrupt changes in the state of the dryland ecosystems, acting as the proverbial "straw that breaks the camel's back" leading to ecosystem degradation. But how ecosystem structure and function change in response to joint changes in aridity and grazing pressure remains a key question. "Quantitative change leads to qualitative change" is a common viewpoint in philosophy, science, and systems thinking, meaning that the gradual accumulation of quantitative changes within a system can ultimately trigger a fundamental transformation in the system. This principle reflects a universal law of development

B. Fu, *The Soul of Geography*, https://doi.org/10.1007/978-981-95-1062-7_3

and change; it is also the basic starting point for us to think about investigating this problem.

Building upon the major project funded by the National Natural Science Foundation of China, we conducted a nearly 4000-km transect survey across the dryland regions of northern China. Integrating multi-source remote sensing data, existing literature, and monitoring data from ecosystem field stations, we developed a novel threshold model that considers the combined pressures of increasing aridity and grazing. This model expands upon existing models that primarily focus on single environmental stressors, such as increasing aridity, and is designed to identify the threshold characteristics of the dryland ecosystems in China under the combined impacts of increasing aridity and grazing.

The study found that as the degree of aridity gradually increases, this "quantitative change" reaches thresholds of 0.7, 0.8, and 0.95, leading to a non-linear decline in vegetation productivity, soil fertility, and plant species richness, respectively, representing a "qualitative change". Specifically, soil organic carbon exhibits a linear decrease with increasing aridity, while soil inorganic carbon abruptly increases. Consequently, total soil carbon displays a trend of initial decrease followed by an increase, with the turning point occurring at an aridity level of 0.71. This compensatory relationship between organic and inorganic carbon contributes to the relative stability of total soil carbon across semi-arid, arid, and hyper-arid zones. Furthermore, grazing often exacerbates the effects of aridity, leading to a premature crossing of the aridity threshold of dryland ecosystems, thereby increasing their susceptibility to "qualitative changes". While the current levels of grazing intensity may appear sustainable in certain areas, future global warming scenarios pose a substantial threat. Under Representative Concentration Pathways (RCPs) 4.5 (medium emission scenario) and 8.5 (high emission scenario), the proportion of such areas suitable for grazing will decrease to 19.7% and 18.3% respectively. The study enriches the theoretical understanding of ecosystem thresholds and provides a scientific basis for the sustainable management of grazing practices in drylands.

Related results were published in *Nature Sustainability* and other publications.

Simple Yet Not Simplistic

The social–ecological system is at the core and frontier of geographical research, characterized by the interaction and feedback mechanisms between social systems and ecological systems. The Loess Plateau is one of the regions in China where the conflicts between population, resources, and environment are most acute, making it an ideal place to study the evolution and sustainability of social–ecological systems.

Supported by the National 973 Program and a key project of National Natural Science Foundation of China, we began in 2018 to guide selected PhD students to engage in research in this new and cutting-edge direction. A social–ecological system is a typical complex system. Research on complex systems explores functional changes through characteristics such as "entropy", "emergence", and "resilience".

Given the thousands of years of human activity on the Loess Plateau, our first focus was the evolution of its social–ecological system over the past millennium. How can we characterize the social–ecological system of the Loess Plateau? My student began by attempting to use multiple factors and indicators to represent it, but due to the lack of historical data and the difficulty of quantitatively reflecting integrated feedback mechanisms, he had to abandon this approach. I jokingly told the student: "You need to seriously study Chairman Mao's essay *On Contradiction*—grasp the principal contradiction, and focus on the principal aspect of the contradiction."

Together, we analyzed and identified that population and the accompanying food demand have been the core issue facing society in the Loess Plateau throughout its long history, and can represent the social system. Forests, through regional processes like the carbon cycle, regional hydrology, and soil erosion, affect the environment and are key components of the ecological system. Cropland, as a form of land use, links the social and ecological systems and plays an important role in regional food security. Thus, a social–ecological system, composed of population–forest–cropland interactions and feedbacks, was formed (see figure below). Based on changes in the interactions among these components, we can identify different evolutionary phases of the social–ecological system and carry out analyses of driving forces and effects, and systematically characterize the features and interaction mechanisms of the Loess Plateau's social–ecological system.

Although it consists of only three elements, the interactions and interdependencies among population, forest, and cropland can effectively reflect the complexity of the social–ecological system—simple, yet not simplistic. By integrating historical records, reconstructed data for historical periods, and observational and statistical data after 1949, we analyzed how the relationships among population, cropland area, and forest coverage have changed over the past millennium on the Loess Plateau, and divided the system's evolution into five phases: ① fast expansion of cultivation (1100–1750s), ② slow expansion of cultivation (1750s–1950s), ③ landscape engineering for higher production (1950s–1970s), ④ transition from cultivation to ecological conservation (1980s–1990s), and ⑤ revegetation for environment (2000s to the present). We also established empirical links between the state of the social–ecological system on the Loess Plateau and its driving forces—such as policy, climate, and socioeconomic factors—as well as local and spillover effects including local food production, sediment load of the Yellow River, natural runoff, delta area, and the frequency of natural levee breaches downstream.

The implications of our findings are significant, especially in highlighting the limitations of single-goal, locally focused approaches to managing the Loess Plateau's social–ecological system. Management practices focused on food security in the first three phases, resulting in local environment degradation and generating negative spillover effects in distant coupled systems. The soil erosion control measures and revegetation in the Loess Plateau reduced soil erosion and sediment load in the Yellow River but contributed to the shift of the Yellow River delta into an erosion state. Furthermore, mismanagement of the planted vegetation in some areas caused formation of a dry soil layer, which had negative influences on both water availability and food production. Therefore, social–ecological system management needs an integrated and systemic perspective, which progresses from pursuing a single goal to considering social–ecological interactions and from focusing on local effects to considering cross-scale effects.

This research was published in *Science Advances* in 2020.

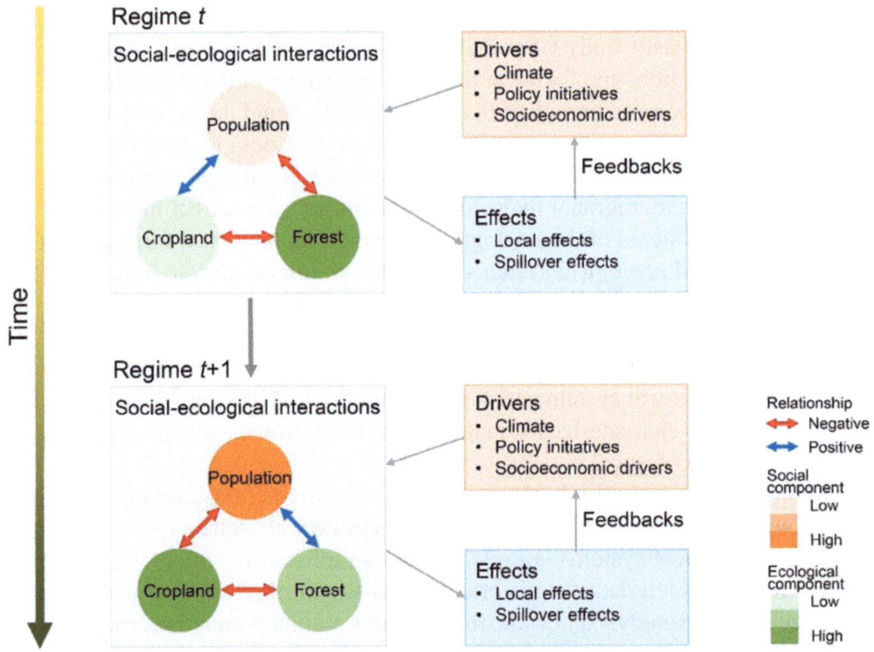

Research framework for identifying the evolutionary phases of social–ecological systems based on changes in system element interactions

Feedback and Effects in the Coupled Human and Natural System

Geography is the science that studies the interactions and spatiotemporal patterns of natural and human elements on the Earth's surface. At its core lies the understanding of human–nature relationships. Humans and nature are intrinsically interconnected. Faced with the increasingly pressing challenges of sustainable development, it is essential to shift from viewing humans and nature as separate systems to viewing them as interacting components of a complex dynamic coupled human and natural system. Such a perspective facilitates a more integrated understanding of the coupling relationship between human activities and the natural environmental processes, rather than just focusing on a single component or one-way relationship.

The Loess Plateau is not only the cradle of ancient civilizations in human history, but also a region characterized by economic underdevelopment and severe ecological degradation in China. This area has long faced prominent development challenges including sparse vegetation coverage, acute soil erosion, intense population pressure, and prevalent livelihood poverty. To address these issues and achieve regional sustainable development, it is crucial to strengthen the coupling analysis of human–natural systems and investigate the interaction mechanisms between human activities and natural environments. Based on identifying the evolutionary phases of the Loess Plateau over the past millennium, our research focused on the most recent phase (2000s to present). We systematically quantified changes in various elements of the coupled human and natural system, analyzed their mutual feedback mechanisms, and conducted in-depth investigations into how human and natural factors dynamically interact within this complex system.

We found that, since 2000, the Loess Plateau has generally achieved a "win–win" outcome in ecological protection and socioeconomic development. Following the launch of the Grain for Green program, the vegetation coverage has increased from 31.6% in 1999 to 67% in 2020, marking a historic transition from "yellow" to "green". After 2000, the vegetation index growth rate on the Loess Plateau surpassed the national average, placing it at the forefront of China's greening efforts. With the restoration of vegetation, the region's carbon storage has significantly increased, transforming its ecosystems from a net carbon source to a carbon sink. Soil and water loss has also been effectively controlled, the amount of sediment entering the Yellow River has significantly reduced. Between 2001 and 2020, average sediment load in the middle reaches of the Yellow River dropped to 240 million tons, comparable to levels recorded over a millennium ago when anthropogenic disturbance was minimal. The agricultural productivity has also significantly improved. From 2000 to 2020, annual grain output increased by 47.05%, exceeding the national average for the same period. Rural income sources have diversified, expanding from traditional agricultural production to non-agricultural employment, orchard management, ecological compensation, and shareholding dividends, thereby optimizing the income structure. Concurrently, socioeconomic development of the Loess Plateau has accelerated, with industrial structures becoming increasingly diversified. The urbanization rate increased from 34.9% in 2000 to 63.8% in 2020, significantly altering the urban–rural population structure and alleviating rural population pressure.

Further analysis showed that the Loess Plateau previously suffered from a social-ecological trap, characterized by the interlinked degradation of ecosystems and livelihood poverty—a cycle difficult to break. This vicious cycle, driven by factors such as the high erodibility of loess soils, intensive cultivation, population pressures, and extreme weather events, has been gradually reversed in the twenty-first century through targeted ecological restoration and development strategies. The region's governance experience offers several key lessons: ① Promoting urbanization and diversifying income structures. This helps to reduce population pressure in rural areas and reduce long-term dependence on agricultural income. ② Implementing engineering interventions tailored to local conditions. Projects such as terraces and check dams, tailored to the Loess Plateau's geomorphological characteristics, have played an important role in effectively controlling soil and water loss and creating high-quality farmland. ③ Investing in ecological restoration. For example, the Grain for Green Program on the Loess Plateau has played a pivotal role in both restoring vegetation and improving local livelihoods.

After identifying the Grain for Green Program as the main influencing factor of ecological improvements and an important contributor to socioeconomic transformation on the Loess Plateau in the twenty-first century, we further focused on the feedback of the social-ecological factors on the Grain for Green Program. We found that there were spatial differences in the effectiveness of afforestation under the Grain for Green Program. The aridity index, rural non-farm employment, and rural migration improved afforestation effectiveness, whereas the total afforestation degree (the cumulative area of afforestation divided by the total area), vegetation conditions before afforestation, grain production, and investment in fixed assets decreased its effectiveness. As for participants' land reconversion willingness after the Grain for Green Program, we found that household income, ecological awareness, and employment changes negatively affected reconversion willingness, while nonfarm employment positively affected it at the individual level. At the regional level, grain production and water yield changes could influence the reconversion willingness of respondents with different individual characteristics.

Despite the notable progress achieved, several challenges remain. In some areas, the structure of artificially restored vegetation remains overly simplified, with limited ecological functionality, and excessive planting densities leads to serious water stress. The fragile environment of the Loess Plateau has not fundamentally altered, some areas have a high risk of natural disasters, and the coordination between the middle and lower reaches is relatively lacking. Addressing these ongoing issues requires an integrated strategy grounded in a deep understanding of coupled human and natural system dynamics. Future efforts should prioritize ecosystem stability and resilience, promote green socioeconomic transitions, improve territorial spatial planning, and enhance coordinated basin-wide governance. In particular, the focus of ecological protection and restoration should shift from large-scale vegetation establishment to the consolidation and functional optimization of existing restoration outcomes. We further conducted in-depth research on *Robinia pseudoacacia*—a widely used tree species in vegetation restoration on the Loess Plateau. Excessive planting density is a major cause of soil desiccation and tree growth decline in the loess hilly and gully areas. Scientific thinning practices can optimize stand structure and regulate soil moisture by controlling vegetation transpiration water consumption, serving as an effective approach to enhance the sustainability of *Robinia pseudoacacia* plantations. Our research elucidated the impacts of different thinning intensities on the growth and water use efficiency of these plantations. Preliminary findings revealed that stands with 52% thinning intensity exhibited higher water use efficiency, providing critical evidence for determining appropriate thinning intensities and optimal stand densities.

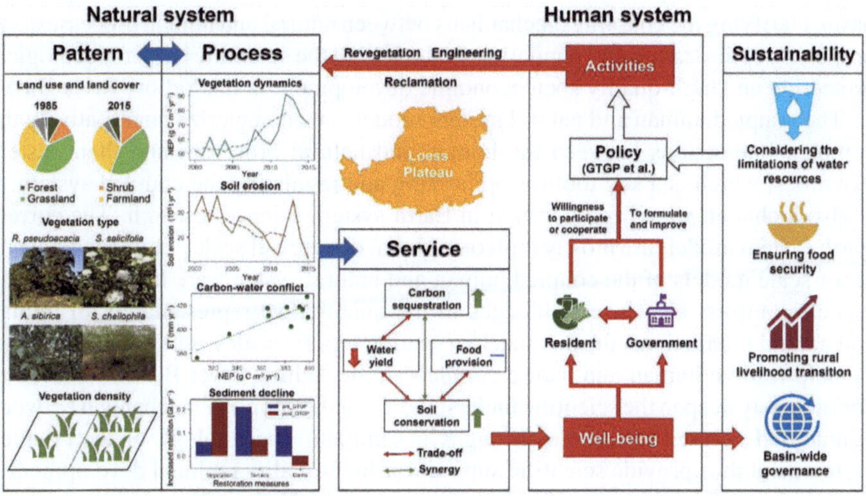

Illustration of the coupled human and natural system studies of China's Loess Plateau using the pattern–process–service–sustainability framework

Modelling the Coupled Human and Natural System

The Yellow River is the mother river of the Chinese nation. "When the Yellow River is calm, the world is at peace". A calm Yellow River has been a long-standing hope for thousands of years. Most of the Yellow River Basin is characterized by arid and semi-arid climates, covering 8.3% of the country's land area. Although the annual runoff of the Yellow River only accounts for 2% of the total in China, its water supply supports 12% of China's population, 15% of the cultivated land, and 13 national energy and chemical industry bases. The conflicts between human demand and water resources in the basin are prominent, with the utilization rate of water resources close to 80%, far exceeding the ecological warning line of 40%. Ecosystem degradation in the upper reaches of the Yellow River, soil and water loss in the middle reaches, and low ecological flow in the lower reaches create multiple ecological pressures on the basin. These issues reflect the Yellow River Basin's disharmony between human and nature. Due to the seriously imbalanced relationship between socioeconomic development and hydrological and ecological processes, the Yellow River Basin is one of China's regions with the most prominent and complex human and natural conflicts.

Coupled human and natural systems constitute a complex framework where the geographical environment and human societies continuously influence one another. Key characteristics of the coupled human and natural systems of the Yellow River Basin include pronounced conflicts between high water demand and limited supply, along with significant regional disparities across the upper, middle, and lower reaches. Therefore, revealing the dynamic evolution of human and natural components in the basin, clarifying the coupling mechanisms between natural and human processes, and proposing optimization and regulation strategies are the scientific basis for ecological protection and high-quality socioeconomic development in the Yellow River Basin.

The coupled human and natural system model can characterize quantitatively the complex interactions between the human and natural processes and their system dynamics, which is a key tool for optimizing and regulating the coupled system. It is also a challenge and frontier area in Earth system science research. The current Earth system models are mostly built on a global or national scale, while regional and basin-scale models of the coupled human and natural systems are lacking. Existing modeling efforts also face challenges in the quantitative representation of human and natural interactions, the mismatch in spatiotemporal scales, data availability, etc. To help resolve human and nature conflicts in the Yellow River Basin, we need to continuously deepen the scientific understanding of the coupling mechanism between human and nature, make breakthroughs in simulation technology of the coupled system, and thus provide scientific support for high-quality regional development.

Based on in-depth knowledge and analyses of the Yellow River Basin, we have adopted a system perspective and designed a model of the coupled human and natural systems in the Yellow River Basin based on system dynamics approaches, with an emphasis on the relationship between human and water in terms of water demand and supply. The model contains ten modules to represent the major human and natural processes in the Yellow River Basin, including population, economy, energy, food, water demand, water supply, land, sediment, carbon, and climate. Each module is built based on physical laws, economic theories or empirical relationships, and the different modules are linked to each other with interactions. The model can simulate the two-way interactions and the influence of the coupled human and natural systems. For example, population changes can affect regional economic development through labor supply; meanwhile, economic growth can also affect population migration and mortality, thereby changing the population; changes in population and economy will simultaneously affect various resource demands for food, water, energy, and land use, etc., thereby affecting runoff, ecosystem productivity, carbon absorption, and other natural processes.

The coupled human and natural systems model of the Yellow River Basin can simulate and predict the complex dynamics of the basin, evaluate changes in essential human and natural processes under different policies and development scenarios—such as food production, ecological protection, water resources supply and demand, water and sediment relationships, etc., and propose optimized paths to ensure ecological, food, economic and social security. Research on the coupled human and natural systems model of the Yellow River Basin promotes the development and application of comprehensive simulation model technology in geography. It provides an important scientific decision-making tool for the basin to achieve harmony between human and nature and high-quality development.

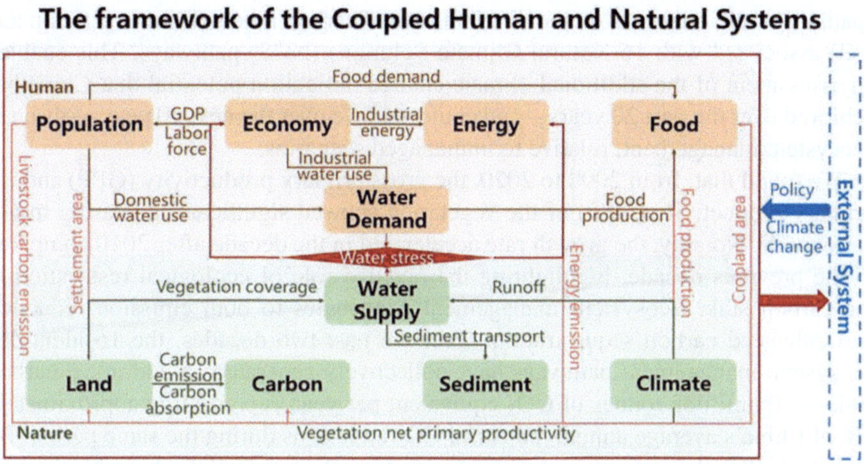

Overview of the coupled human and natural system model framework in the Yellow River Basin

"Managing" for Benefits

In the context of the urgent need to address global warming, terrestrial ecosystem management strategies that aim to increase carbon sinks and reduce greenhouse gas emissions, while also improving the living environment, enhancing ecosystem resilience, and improving human well-being, are defined as "Natural Climate Solutions" (NCS).

Terrestrial ecosystems play a crucial role in carbon sequestration and serve as an important means of mitigating climate change. However, there remains considerable uncertainty in estimating ecosystem carbon sequestration at the national and regional scales. Over the past 20 years, China has achieved remarkable ecological and environmental outcomes by implementing projects such as the Natural Forest Protection Program, the Grain for Green Program, and the Returning Grazing Land to Grassland Program. Large areas of secondary forests and plantations have been protected and rebuilt, and degraded grasslands and wetlands have been effectively rehabilitated. However, the specific contribution of China's ecosystem management to carbon sequestration, particularly in the context of climate change and human activities, remains unclear. Supported by the National Key Research and Development Program "Impacts of Global Change on Ecosystem Services", we established for the first time an accounting framework for China's NCS. The framework encompasses 16 pathways of protecting, restoring, and managing forests, grasslands, wetlands, and farmland ecosystems. By integrating active and passive microwave remote sensing, combined with machine learning methods, the vegetation carbon storage dataset (2000s–2020s) has been improved and generated. By integrating ecosystem inventory data with literature sources, public databases, and policy documents, we developed spatially explicit datasets on the distribution and carbon fluxes (from vegetation and soil) associated with 16 Natural Climate Solutions (NCS) pathways. This enabled an assessment of the additional climate change mitigation potential that China has achieved over the past 20 years—and could achieve over the next 40 years—through ecosystem management, relative to unmanaged scenarios.

We found that, from 2000 to 2020, the gross primary productivity (GPP) and net primary productivity (NPP) of the vegetation showed significant increasing trends nationwide. Notably, the growth rate accelerated in the decade after 2010 compared to the previous decade, highlighting the positive role of ecological restoration in the carbon sink. Ecosystem management contributes to both emission reduction and enhanced carbon sequestration. Over the past two decades, the 16 identified ecosystem management pathways have collectively generated an additional carbon sink of 600 million tonnes of CO_2 equivalent per year, representing approximately 8% of China's average annual industrial CO_2 emissions during the same period. By 2030 and 2060, the maximum climate change mitigation potential achievable through NCS pathways are expected to be 600 million tons (CO_2 e/yr) and 1 billion tons (CO_2 e/yr), respectively. The four provinces of Inner Mongolia, Heilongjiang, Sichuan, and Yunnan lead in both historical achievement and future potential in this regard. Nationwide, the management of forest ecosystems contributes most significantly to climate mitigation capacity and potential (70%), followed by croplands (16%), grasslands (10%), and wetlands (4%). Before 2020, the primary types of ecosystem management were restoration and improved management. In the coming decades, the scope for ecological restoration will gradually diminish, and the carbon sequestration potential of ecosystems will increasingly rely on a deeper exploration of improved management and conservation measures.

This work presents a systematic approach, having not only established an accounting indicator and methodological framework for NCS at the national scale, but also generated a series of datasets and completed assessments spanning the past 20 years and the next 40 years. Related findings have been published in internationally renowned journals such as *Nature Climate Change*, with the appendix extending hundreds of pages. *Nature Climate Change* published a commentary in the same issue, with the editorial team remarking on this paper: "This paper provides national level estimates of carbon uptake and greenhouse gas emissions reduction potential through managing natural systems (called natural climate solutions) for China. The work complements global-scale analyses with more nuanced regional detail, which is important given the global interest in China reaching its climate mitigation goals." This work were also selected as one of China's Top 10 Ecological and Environmental Scientific Advances in 2022.

Responding to a Changing World

To tackle the most pressing issues facing humanity, such as climate change, poverty, inequality and quality education, the United Nations adopted 17 ambitious Sustainable Development Goals (SDGs) to stimulate actions in critically important areas for people, the planet and prosperity. The 17 SDGs are integrated and indivisible, balancing the economic, social and environmental dimensions of sustainable development. They cover all aspects of human life and interact in complex ways. Actions for one goal may reinforce or offset the actions for another, resulting in synergies and trade-offs among the SDGs. Given the 'leave no one behind' objective of the 2030 Agenda for Sustainable Development, understanding interactions among the SDGs is crucial when designing appropriate and efficient policies to implement them.

Everything in the world is constantly evolving, nothing remains unchanged forever. Do SDG interactions change as sustainable development progresses and, if so, how? To address this question, we used a correlational network approach and a global SDG database of 166 countries to analyze the evolution of SDG interactions along a progression of sustainable development measured by the SDG Index.

The study found that the 17 SDGs first experienced a decoupling followed by a recoupling process as sustainable development level increased: SDGs were both more positively and more negatively connected at low and high sustainable development levels, but they were clustered into more isolated positive connection groups at middle levels.

The dominant SDGs in the interaction network also change with sustainable development levels. In the synergy networks, SDG 3 (good health and well-being), SDG 9 (industry, innovation and infrastructure), SDG 16 (peace, justice and strong institutions) and SDG 6 (clean water and sanitation) were the most prominent SDGs along the index. SDG 4 (quality education) was a dominant SDG when the SDG Index was low but became less important in the synergy networks when the SDG Index was high, and SDG 1 (no poverty) and SDG 7 (affordable and clean energy) showed similar trends. SDG 8 (decent work and economic growth) was less important when the SDG Index was low but became one of the top five dominant nodes when the SDG Index was high. A similar trend was observed for SDG 5 (gender equality). SDG 12 (responsible consumption and production) and SDG 13 (climate action) played dominant roles in the negative interactions among SDGs, especially at high sustainable development levels. SDG 15 (life on land) and SDG 14 (life below water) were also relatively dominant SDGs in the trade-off networks when the SDG Index was low, but they became less important when the SDG Index was high.

The study also analyzed how the SDGs that tend to be achieved together change with sustainable development levels. Based on the positive connections among goals, the 17 SDGs can be clustered into three modules at low sustainable development levels, reflecting the social and economic SDGs, environmental SDGs and partnerships for the goals (SDG 17) respectively. As the level of sustainable development improved, these modules were disrupted, and the number of modules increased as the modules became smaller. At high levels of sustainable development, 13 of the SDGs regroup, leaving only SDG 14, SDG 17 and a pair of SDGs (12 and 13) isolated from these connected SDGs. This also reflected a process of decoupling followed by re-coupling of SDGs.

By revealing for the first time the dynamic changes of SDG interactions along sustainable development levels, the research strengthens our understanding of sustainable development and may help to suggest specific action priorities to achieve as many SDGs as possible by 2030. This study proves the necessity of research about dynamic changes in SDG interactions and may also begin to lay a foundation for analyzing such dynamic changes at different scales.

Related results were published in *Nature Sustainability* in 2022 and were selected as a Featured Article in that issue. In the same issue, *Nature Sustainability* published a two-page commentary by a scholar from Australia, which elaborated on the scientific significance of this work. The commentary stated that "In the global transition to sustainable development, policy coherence is key. A new study highlights the equal importance of policy dynamism over time—how policymakers must anticipate national advances in sustainability to meet shifting strategic needs."

Social Applications

Geography, vast and practical, influences the economy, environment, society, and daily life, in every place and at all times. Through a multidimensional perspective, integrated essence, and systems thinking, geography enables a deeper understanding of the world and offers solutions to improve it, benefiting humanity and enriching individual lives. Theories and methodologies in geography offer unparalleled advantages in addressing the challenges humanity faces regarding resource management, environmental sustainability, and socioeconomic development.

Innovating Geographical Science to Support Ecological Civilization

The realization of a Beautiful China cannot be achieved without strong support from scientific and technological innovation. Aiming to build a Beautiful China, innovating geographical theories, methods, and technologies is the only path toward becoming a leading country in geography and an essential requirement for building a world power in science and technology. Geography is a discipline that studies the spatial distribution patterns, temporal evolution processes, and regional characteristics of geographical elements and geographical complexes. In recent years, the theories, methods, and technologies of geography have been continuously innovated and developed, playing an important role in scientific decision-making. Chinese geography is gradually forming a geographical science system with distinct Chinese characteristics and far-reaching international influence, serving the coordinated and sustainable development of the national economy, society, and environment systems. Ecological civilization, as a new stage in the development of human civilization, is guided by the vision of harmony between humans and nature and is underpinned by the principle of sustainable development. It emphasizes the need to maintain the integrity of natural ecosystems in the course of production and daily life, to preserve

B. Fu, *The Soul of Geography*, https://doi.org/10.1007/978-981-95-1062-7_4

the harmony between humans and nature, and to promote balanced and coordinated relationships among the natural, ecological, economic, and social systems. In the face of numerous resource and environmental problems brought about by rapid population growth, economic development, and globalization, Chinese geography is deeply rooted in the country's major needs, and is innovating and developing comprehensive theories, methods, and technical approaches to support scientific decision-making for ecological civilization construction. Innovating geographical science to support the construction of ecological civilization has become a key task of contemporary Chinese geography.

Development and Innovation of Geographical Science

Geography takes the earth's surface as its object of study and uses a comprehensive, multi-dimensional perspective to investigate the mechanisms of human-natural interactions. Drawing on its disciplinary characteristics of integration, interdisciplinarity, and regionality, geography takes the resolution of complex problems related to resources, the environment, and development as its mission. Geographical research focuses on human–nature relationships, emphasizes the integration of natural and human elements, and aims not only to explain the past, but also to serve the present and predict the future. At present, geography, which has been primarily qualitative, is undergoing a significant transformation toward a more quantitative geographical science. New technological changes, new driving mechanisms, and new research paradigms are driving the continuous deepening and innovation of geographical science in the new era.

Technological changes have provided new technical approaches and new data sources for geographical science. From qualitative description to field experiments, indoor and outdoor simulations, and regional synthesis, geographical science has gradually developed a research pathway centered on the "concept–model–decision" process. This research process begins with integrative geographical thinking, through which conceptual models are constructed and association rules are identified. These rules are then represented in data form and used to express quantitative mechanisms underlying physical, chemical, biological, and human processes. Through interdisciplinary quantification, data verification and calibration are performed, leading to the formulation of mechanistic laws. Finally, through geographical integrative thinking, these laws are synthesized to predict the future and applied in decision-making. At present, with the support of massive spatial data of various types and levels of precision, new geographic information technologies have been integrated into all stages of solving geographical problems. Big data, virtualization, and visualization have become important tools for characterizing complex human and natural processes, conducting scenario simulations, and supporting decision-making in geography. These tools have effectively supported the further deepening of understanding of the processes and mechanisms underlying geographical phenomena.

Global change and sustainable development are driving the emergence of new development goals and focal research themes in geographical science. Climate change, ecological degradation, and environmental pollution are profoundly altering the Earth's environment. To achieve the goals of sustainable development and respond to global environmental change, the overall development trend of geographical science is shifting from diversity to system integration. In the new era, geographical science emphasizes research on the Earth's surface system, particularly the terrestrial surface system, to analyze and understand major challenges facing human society in the context of global change. Therefore, the main characteristics of geographical science development in the new era are reflected in the integration and deepening of geographical processes, the integration of terrestrial surface systems, land–sea interactions, and the application of regional environment management. Using "pattern–process–service–sustainability" as the research linkage, geographical science analyzes the driving mechanisms of global change and explores innovative pathways toward sustainable development, further highlighting the effective linkage between geographical knowledge, science, and decision-making.

The simulation of complex human–natural systems is promoting the formation of a new research paradigm in geographical science. The transition from descriptive geographical knowledge to coupling of patterns and processes represents an effective approach to understanding and addressing integrated geographical problems from a mechanistic perspective. In this framework, "pattern" refers to geographical spatial distribution, "process" refers to spatiotemporal evolution, and "coupling" describes the feedback relationship in which patterns influence processes and processes, in turn, alter patterns. Model simulation is not only a key means for coupling pattern and process, but also an important tool for scientific decision-making through scenario prediction. With the continued enhancement of computer hardware, further innovations in technologies such as remote sensing and geographic information systems, and a deeper understanding of pattern–process coupling, the capacity to simulate complex human–natural systems will gradually improve. Facing the research paradigm of human–natural system simulation, future geographical research will inevitably become more integrated, more forward-looking, and more quantitative.

The Construction of Ecological Civilization Is a Major Strategic Requirement for the Country

Ecological civilization is the sum of the material, spiritual, and institutional achievements that humanity has attained in the process of protecting and building a beautiful environment. It is a systematic undertaking that runs through the entire process and all aspects of economic, political, cultural, and social development, and reflects the level of a society's civilizational progress. Accelerating the reform of the ecological civilization system and building a Beautiful China specifically involves four key tasks: promoting green development, addressing prominent environmental problems, strengthening ecosystem protection, and reforming the ecological and environmental governance system. We must firmly establish the socialist outlook on ecological civilization, promote a new pattern of modernization featuring harmony between humanity and nature, and contribute our generation's efforts to ecological and environmental protection.

Promoting green development requires establishing and improving an economic system characterized by green, low-carbon, and circular economic development. This includes optimizing the industrial layout, upgrading traditional industries, promoting energy conservation and emission reduction, developing a circular economy, adjusting the energy structure, expanding industries related to energy conservation and environmental protection, clean production, and clean energy, as well as advancing national water-saving initiatives and other comprehensive approaches to resource conservation and recycling. On the supply side, green development requires reducing energy and material consumption and achieving a closed-loop connection between the production system and the living system. On the consumption side, green development calls for advocating a simple, moderate, green, and low-carbon lifestyle.

To effectively address prominent environmental problems, it is necessary to uphold the principles of whole-of-society participation and source prevention in key areas such as water, soil, air, and solid waste. Specific measures include accelerating water pollution prevention and control, continuing air pollution control actions, strengthening the control and remediation of soil pollution, enhancing the prevention and control of agricultural non-point source pollution, and improving the treatment of solid and domestic waste, among others. At the same time, at broader regional, national, and global scales, it is necessary to implement comprehensive management of river basin environments and nearshore marine areas, and to carry out initiatives to improve rural human settlements. Efforts should also be made to raise emission standards, actively participate in global environmental governance, and fulfill emission reduction commitments.

Strengthening ecosystem protection needs a particular emphasis on the integrated management of natural resources including mountains, rivers, forests, farmlands, lakes, and grasslands. By implementing major projects for the protection and restoration of important ecosystems, it is necessary to optimize the ecological security barrier system, connect ecological corridors and biodiversity conservation networks, and enhance the quality and stability of ecosystems. Based on the spatial planning concept of "multiple plan coordination", it is necessary to complete the delineation and integration of three control lines: the ecological conservation red line, the permanent basic farmland protection red line, and the urban development boundary.

In the implementation of land greening initiatives, efforts should be made to advance the comprehensive management of desertification, rocky desertification, and soil erosion, strengthen wetland protection and restoration, and improve the prevention and control of geological hazards to reduce disaster risk. At the institutional level, it is necessary to improve the natural forest protection system and strictly uphold the protection of cultivated land. The scope of converting farmland to forests and grasslands should be expanded, along with the expansion of pilot programs for crop rotation and fallow. It is also important to improve conservation and recuperation systems for cultivated land, grasslands, forests, rivers, and lakes, and to establish a market-oriented and diversified ecological compensation mechanism.

Reform of the ecological and environmental supervision system is reflected in strengthening the top-level design and organizational leadership for ecological civilization development, and in resolutely preventing and penalizing behaviors that damage the environment. With the establishment of state-owned natural resource asset management and ecological supervision institutions, the ecological and environmental management system has been further improved. The responsibilities of natural resource asset ownership, land use regulation across all territorial spaces, and ecological protection and restoration are now exercised in a unified manner. This is specifically reflected in the establishment of a land space development and protection system, the improvement of supporting policies for major functional zones, and the creation of a natural protected area system with national parks as its core.

The Support of Geographic Science for the Construction of Ecological Civilization

The practical problems involved in the construction of ecological civilization often exhibit characteristics such as the complexity of elements, the regional specificity of targets, and the interdisciplinary nature of knowledge systems. By taking an integrated perspective on issues related to resources, the environment, and sustainable development, geographical science holds a natural disciplinary advantage in supporting the construction of ecological civilization.

Large-scale geographic data processing and visualization have provided strong data and technical support for the simulation of complex human–natural systems, driving a shift in geographic research paradigms from pattern–process coupling toward full-system human and nature simulations. Aligned with the major national needs of ecological civilization construction, the innovation of geographical science theories and methods is reflected in the further integration of multi-source elements and deeper coupling of multiple processes. This involves detailed characterization of the spatiotemporal relationships among elements within coupled human and natural systems, providing a foundational basis for integrated natural resource management. It also includes monitoring and issuing early warnings about system carrying capacity, which can inform policy improvements for major functional zones and support the establishment of land use and protection systems. In addition, constructing ecological security patterns enables a systematic response to ecological risks at both global and national scales, offering practical pathways for optimizing ecological security barriers and the spatial configuration of mountains, rivers, forests, farmlands, lakes, and grasslands. Furthermore, geographic science supports the analysis of the comprehensive impact of human activities on climate change through direct observation and model simulation, contributing directly to China's participation in global environmental governance and the development of emission reduction measures. It also facilitates the exploration of integrated sustainable use scenarios for food, energy, and water, offering regionally practical solutions for comprehensive resource conservation and recycling. By linking biodiversity conservation with ecosystem service management, geographic science enhances the quality and stability of regional ecosystems from a systems perspective and supports the establishment of diversified ecological compensation mechanisms. In addition, it contributes to the assessment of the health impacts of environmental pollution, helping address pressing environmental issues related to water, soil, air, and solid waste while safeguarding the health of urban and rural residents. Lastly, it explores sustainable pathways for urbanization and rural revitalization by optimizing the spatial distribution of industries and improving living environments, thereby promoting a new model of modernization that advances harmony between humanity and nature.

China's vast territory has provided an excellent context and experimental ground for the study of human–nature relationships, while its large population base and rapid economic growth have posed unprecedented challenges and also created unprecedented opportunities for such research. Upholding the historical tradition of promoting disciplinary development through task orientation, Chinese geography has continued to evolve and improve by serving governmental decision-making and national needs. The construction of ecological civilization represents a major historical opportunity for integrating Chinese geographical science with national decision-making. Promoting green development, addressing prominent environmental problems, strengthening ecosystem protection, and reforming the ecological and environmental supervision system are key areas of focus in contemporary geographical science. With the goal of building a Beautiful China, and grounded in the frontiers of geographical science research, efforts to innovate the theories and technical methods of geographical science will further advance China's transformation from a major geographical nation to a geographical powerhouse, and more effectively demonstrate the important disciplinary value of geography in supporting scientific decision-making.

Originally published in Chinese in *A Hundred Academicians Talk about Building a Scientific and Technological Powerhouse*, 2019

Current Conditions, Issues, and Suggestions for Ecological Protection and High-Quality Development in the Loess Plateau

The Loess Plateau, located in the mid-northern region of China and the middle reaches of the Yellow River, covers an area of approximately 640,000 km^2. It is recognized as an important cradle of Chinese civilization. However, the region is ecologically fragile and highly sensitive to climate change. It is also one of the areas in China where conflicts among population growth, resource limitations, and environmental degradation are most pronounced. For a long time, the development of the Loess Plateau has been hindered by severe soil erosion, sparse vegetation cover, high population pressure, and low land productivity. In response, a series of ecological restoration and land management initiatives—such as slope management, integrated slope-gully management, small watershed comprehensive management, the Grain for Green Program, and gully land consolidation—have been implemented. Notably, since the launch of the Grain for Green Program in 1999, the Loess Plateau has witnessed substantial ecological improvement. These efforts have led to a "win–win" situation for environmental protection and socioeconomic development. The Loess Plateau has thus become a flagship example of China's ecological civilization construction and offers a valuable reference for ecological restoration efforts worldwide.

The "Yellow River Basin's Ecological Protection and High-quality Development Plan", issued by the Central Committee of the Communist Party of China and the State Council, emphasizes the strategic importance of the Loess Plateau. The document calls for prioritizing soil and water conservation, fully protecting natural forests, consolidating the gains of the Grain for Green Program, intensifying the comprehensive soil and water loss control, promoting steady urbanization, and improving the ecological landscape of the middle reaches of the Yellow River. With the advancement of the national strategy for ecological protection and high-quality development of the Yellow River Basin, the ecological construction of the Loess Plateau has entered a critical phase. This new stage is characterized by the consolidation of ecological governance outcomes and the transition toward integrated socioeconomic development. This chapter systematically examines the current status and major challenges related to ecological and socioeconomic development in the Loess Plateau, and proposes targeted strategies and policy recommendations to promote ecological protection and support high-quality development in the Loess Plateau during this new era.

Current Conditions of Ecological Construction and Socioeconomic Development in the Loess Plateau

Through years of sustained ecological restoration and development efforts, the environment of the Loess Plateau has shown marked improvement, accompanied by substantial progress in socioeconomic development.

I. Significant increase in forest and grassland coverage and enhancement of ecosystem carbon sequestration capacity

Since the implementation of the Grain for Green program, vegetation coverage in the Loess Plateau has increased from 31.6% in 1999 to 67% in 2020, achieving a historic transition from "yellow" to "green". After 2000, the region's vegetation index has grown at a rate exceeding the national average, positioning the Loess Plateau as a leader in national greening efforts. As vegetation has recovered, the carbon sequestration of the Loess Plateau has significantly improved, shifting its ecosystems from a carbon source to a carbon sink. The increase in carbon sequestration is mainly concentrated in the hilly and gully areas, where reforestation initiatives have been extensively implemented.

II. Effective control of soil erosion and substantial reduction in sediment load of the Yellow River

According to the Soil and Water Conservation Bulletin in the Yellow River Basin, by the end of 2020, a total area of 252,400 km^2 of soil erosion had been effectively managed, and 58,100 check dams had been constructed. The soil and water conservation rate in the Yellow River Basin increased from 41.49% in 1990 to 66.94% in 2020, with the rate in the Loess Plateau reaching 63.44%. Data from the Tongguan hydrological station show that the average annual sediment transport in the middle reaches of the Yellow River dropped to 240 million tons between 2001 and 2020—comparable to levels recorded over 1,000 years ago when human activities were rare.

III. Enhanced agricultural productivity and improved rural income structure

By promoting the construction of high-standard farmland through engineering measures such as dam construction and gully land consolidation, the agricultural productivity of the Loess Plateau has significantly increased. From 2000 to 2020, the total grain output increased by 47.05%, outpacing the national average growth rate of 44.86% during the same period. Studies indicate that the grain self-sufficiency rate in the Loess Plateau increased from 62% in 1950 to 108% in 2019. Rural income sources have diversified, expanding from traditional agriculture to non-agricultural employment, orchard management, ecological compensation, and shareholding dividend. As a result, the income structure has been optimized, with the *per capita* net income of rural residents increasing from 1,916 yuan in 2000 to 14,400 yuan in 2020. By 2020, all registered poor populations in the region had been lifted out of poverty, and all poverty-stricken counties had been officially delisted, achieving historic achievements in poverty alleviation and the elimination of absolute poverty.

IV. Accelerated socioeconomic development with diversified industrial structures

The urbanization rate in the Loess Plateau increased from 34.9% in 2000 to 63.8% in 2020, resulting in a significant shift in the urban–rural population structure and alleviating population pressure in rural areas. The scale of the economy and society expanded rapidly, with key economic indicators outperforming national averages. Between 2000 and 2020, the region's Gross Domestic Product (GDP), total industrial output, and *per capita* GDP increased by 14.35, 10.86, and 11.79 times, surpassing national growth rates of 10.11, 7.77, and 9.04 times respectively. The region's share of total national industrial output increased from 6.11% in 2000 to 8.54% in 2020.

Some areas have leveraged local resources and locational advantages to develop specialized agricultural and animal husbandry industries such as apples orchards, jujubes and Sichuan pepper cultivation, and livestock farming such as sheep and beef cattle. Additionally, ecological tourism and red tourism have emerged as new growth engines, promoting rural revitalization and integrated urban–rural development.

Key Issues in the Ecological Construction and Socioeconomic Development of the Loess Plateau

Despite notable achievements in ecological restoration, the Loess Plateau still faces several challenges, including limited sustainability of artificial vegetation, fragile environment, rural depopulation, and uneven regional development.

I. Irrational vegetation restoration design and limited ecological functionality

(1) Oversimplified vegetation structure and limited ecological performance

To improve the seedling survival rates and accelerate greening, afforestation efforts in the Loess Plateau have relied predominantly on fast-growing and low-cost species, such as *Robinia pseudoacacia* and poplar. In many areas, plantations consist of only two or three tree species planted in fixed ratios (e.g., 1:1 or 2:1), resulting in structurally simplified forests compared to natural secondary forests. These artificial forests perform poorly in terms of biodiversity and ecological functionality. Monitoring studies indicate that the biodiversity and ecological function of artificial forests are 10–15% and 19–65% lower than those of natural secondary forest reference sites, respectively. In addition, the simplified vegetation structure limits biotic resistance to herbivorous insect outbreaks, which in turn compromises the physiological function and stability of the vegetation community.

(2) Excessive planting density and critical water resource depletion

The Loess Plateau has deep soil layers, and vegetation growth mainly depends on precipitation and soil moisture. High initial planting density of artificially restored vegetation have led to excessive soil water consumption, expansion of dry soil layers, and widespread "little old man trees" or dieback of trees. Research shows that vegetation coverage in the central and northern regions of the Loess Plateau has exceeded the climate-defined equilibrium vegetation coverage, and high vegetation density is the primary driver of soil drying. Current vegetation coverage is approaching the threshold of vegetation carrying capacity, and exceeding this threshold will cause water use conflicts between ecological restoration and socioeconomic development.

II. High risk of natural disasters in some areas, poor coordination in the middle and lower reaches of the Yellow River

(1) Persistently fragile ecosystems and high disaster risks

The ecological fragility of the Loess Plateau remains fundamentally unchanged, and the region is highly susceptible to natural disasters. The Loess Plateau is located in a typical temperate continental monsoon climate zone, with frequent extreme climate events such as heavy rain, drought, and frost. These hazards threaten the stability and sustainability of gully agriculture and specialty fruit production. Field surveys have found that, under heavy to extreme heavy rainfall, the erosion risk remains high, leading to downstream sedimentation and flooding. The check dams play a significant role in reducing sediment transport, many are vulnerable during extreme weather event. Notably, 54.54% of the check dams in the Loess Plateau were constructed before 1986, mostly as small- to medium-sized dams with low engineering standards, most of which are now filled with sediment, in disrepair, or long neglected. Of the 5,282 identified high-risk dams, 456 have residential areas downstream. The likelihood of dam failure is rising due to increasing climate extremes and dam aging.

(2) Altered water–sediment dynamics and persistent uncoordinated water–sediment relationship in the Yellow River

The shortage of water resources and the uncoordinated relationship between water and sediment are important bottlenecks for ecological protection and high-quality development in the Yellow River Basin. The Loess Plateau is a major source of sediment but a region with weak runoff generation in the Yellow River Basin. With the large-scale ecological management of the Loess Plateau, regional runoff has significantly decreased. For example, from 2000 to 2016, despite increased precipitation, the annual average runoff in the Hekou–Longmen section of the Yellow River decreased by over 40% compared to 1980–1999. While sediment yield has declined and downstream sedimentation has lessened, this has led to severe riverbed scouring. From 1987 to 2020, frequent shifts in the Yellow River channel between Huayuankou and Liuyuankou has caused the loss of about 32 km^2 of farmland due to erosion.

III. Persistent rural vulnerability and regional development gaps

(1) Low land-use efficiency and fragile rural livelihoods

With urbanization and socioeconomic development, a large number of young and middle-aged rural laborers have moved to cities for work, accelerating rural depopulation and aging. In 2020, the aging rate of the rural population in the provinces across the Loess Plateau reached 14.16%, higher than the national average of 13.52%. Permanent rural residents accounted for only 36.2% of the registered population. The aging of the rural population has hindered the development of intensive farming, and different degrees of abandonment of farmland have appeared in rural areas. As national subsidies policies for reforestation and poverty alleviation phase out, the income structure that previously relied on transfer payments faces new challenges. Although rural *per capita* net income continues to grow, it remains below the national average, and the urban–rural income gap continues to widen—highlighting the ongoing challenges in achieving common prosperity.

(2) Significant regional development disparities and underdeveloped county-level economies

The industrial structure in the Loess Plateau is predominantly resource-dependent. While counties rich in natural resources rely heavily on resource-based industries, other counties have seen sluggish development. The Gini coefficient of the added value of secondary industry has increased from 0.34 in 2000 to 0.49 in 2020, indicating widening disparities. More than 67.74% of the districts and counties in the region have *per capita* GDP and fiscal revenue lower than 50% of the regional average. Since 2010, the proportion of secondary industry has declined due to resource depletion and production capacity adjustments, while growth in tertiary industry remains slow. Fiscal revenue growth lags behind spending, leading to chronic deficits and underinvestment in county-level infrastructure and development. The outflow of population is serious, which restricts the development of the county economy. 86.22% of counties faced fiscal deficits, with the regional fiscal shortfall increasing from 10.79 billion yuan in 2000 to 638.83 billion yuan in 2020. A total of 44.57% of counties had deficits exceeding 2 billion yuan. Limited job opportunities and inadequate public services further hinder rural revitalization and economic transformation.

(3) Underutilization of ecological resources and insufficient value realization of ecological products

The improved environment of the Loess Plateau plays a key role in reducing sediment inflow into the Yellow River and protecting the lives and property of downstream people. However, the valuation and compensation mechanisms for ecosystem services are still at an early stage. Although the compensation standard for the Grain for Green Program has been adjusted from 230 yuan per mu per year (for 5–8 years) to 1500–1600 yuan per mu in the latest policy round (5 years), this is still low compared with the opportunity costs of off-farm employment and the true value of ecosystem services. While ecological improvements have laid a solid foundation for ecotourism and agriculture in the region, mechanisms for translating ecological assets into economic value—i.e., implementing the Two Mountains theory and realizing ecological industrialization—are still under development. A comprehensive, market-oriented mechanism for cultivating and monetizing ecological products has yet to be established.

Suggestions for Ecological Protection and High-Quality Development in the Loess Plateau

In light to the aforementioned issues, it is imperative to develop feasible countermeasures and long-term mechanisms in the analytical framework of the coupled human and natural system. These efforts should focus on improving ecosystem stability and sustainability, promoting a green transition in socioeconomic development, implementing scientific territorial planning, and adopting an integrated and coordinated approach to managing the entire Yellow River Basin during ecological restoration efforts. Such integrated approaches are essential for advancing ecological protection and high-quality development in the Loess Plateau, ultimately facilitating coordinated ecological–economic development and harmonious human–nature coexistence.

I. Improving ecosystem stability and sustainability for carbon neutrality

(1) Maintaining a reasonable vegetation density

Vegetation restoration in the Loess Plateau is close to the limit of vegetation carrying capacity. Future ecological protection and restoration should shift its focus from extensive afforestation to consolidating achievements and enhancing ecological functions. High forest density, particularly in *Robinia pseudoacacia* plantations in hilly and gully areas, is a key factor contributing to soil drying and forest decline. Scientific thinning and structural optimization of high water-consuming monoculture forests are necessary to reduce evapotranspiration and regulate soil moisture, thereby improving sustainability.

(2) Establishing environmentally adapted vegetation communities

Promoting adaptive ecosystem management by conducting comprehensive assessments of environmental suitability, soil and water conservation benefits, socioeconomic outcomes, and ecological risks associated with existing restoration measures. Targeting artificial forests with low adaptability and high ecological risk (such as high water consumption, pests and diseases) for restructuring through near-natural transformation. Guided by zonal vegetation patterns and successional dynamics, introducing native and drought-tolerant species such as Chinese juniper and sand willow to gradually replace less suitable high water-consuming species and enhance biodiversity and ecological resilience.

(3) Developing a market for ecological carbon sink

With the increase in forest and grass coverage, the carbon sink capacity of the Loess Plateau's ecosystems has improved. Establishing a carbon sink trading system and launching pilot projects can facilitate the monetization of ecosystem services. Strengthening the carbon measurement and monitoring framework, incentivizing social capital investment, and supporting local forest farmers' participation in carbon markets can promote sustainable forest and grassland management while contributing to national carbon neutrality and rural revitalization goals.

II. Promoting a green transition in socioeconomic development for rural revitalization

(1) Industrializing ecological resources and realizing value transformation

Gradually incorporating natural assets and ecosystem services into the national economic accounting system, and establishing a comprehensive accounting system for ecosystem production. According to the logic of ecological resource cultivation, assetization, capitalization, and marketization, comprehensively utilizing land engineering technology, effectively improving the quality and function of regional ecological resources. Reforming property rights and benefit-sharing mechanisms, nurturing new business entities and specialized markets, and facilitating the two-way flow of rural ecological resources and urban capital to achieve value realization.

(2) Developing a green, low-carbon, circular economy and promoting resource-saving and intensive use

Leveraging integrated agricultural systems including internal cycles (framing-breeding), external cycles (agricultural processing and e-commerce) and the multifunctional large cycles (agriculture and tourism, health care, and rural–urban integration). Establishing multifunctional industrial chains to drive mutual reinforcement between regional and basin-wide development. Through institutional innovation and multi-actor collaboration, ensuring the efficient utilization of local resources and promoting coordinated human–nature development.

(3) Using information technology to promote diversified and informatized management of industries

Capitalizing on the resource endowments and cultural heritage of the Loess Plateau by leveraging information technologies for efficient resource allocation and evidence-based decision-making. Promoting emerging rural industries such as organic farming, boutique agriculture, and smart agriculture. Supporting the development of modern agricultural models including park-based, mountain-based, and channel-based systems to revitalize rural areas and foster smart villages.

III. Implementing scientific territorial planning for integrated conservation and restoration

(1) Adjusting and optimizing living space

Based on geographical patterns, population and economic distribution, and urbanization stages, delineating the "production–living–ecological" spatial structure and pattern. Scientifically promoting village classification, optimizing and adjusting the urban–rural system, and strengthening county-level hubs. Promoting comprehensive land consolidation, introducing market mechanisms, improving public resource allocation, and upgrading the rural living environment.

(2) Efficiently using production space

Aligning with ecological civilization and urban–rural coordination strategies, prioritizing land optimization and reforming financial mechanisms to mobilize new agricultural business entities. Scientifically promoting the organization, mechanization, efficiency of agriculture. Scaling up multifunctional agriculture and integrating farming with education, research, tourism, and healthcare to establish integrated rural ecological complexes.

(3) Protecting and restoring ecological space

Establishing an ecosystem health assessment and governance framework that emphasizes holistic restoration. Employing biological, engineering, and administrative measures for small watershed management and check dam construction. Consolidating soil and water conservation achievements and strengthening the Loess Plateau's ecological security barrier.

IV. Establishing monitoring and basin-wide governance systems for soil and water conservation

(1) Implementing monitoring systems for soil and water conservation

Slope erosion and gully-induced sediment are still major contributors to Yellow River sediment load. Establishing a dynamic monitoring system centered on observation stations and supported by regular surveys. Improving early warning capabilities and target interventions in high-risk areas. Reinforcing vulnerable terraces and check dams, maintaining reclaimed land projects, and enhancing the resilience of ecological infrastructure to extreme climate events (such as heavy rainstorms).

(2) Strengthening coordinated decision-making

Integrating scientific data and building an intelligent decision support system for ecological construction in the Loess Plateau. Connecting project supervision with governance effectiveness evaluation, and facilitating data-driven forecasting, early warning, and management. Introducing third-party evaluation mechanisms and building a science–technology–policy interface to support evidence-based and interdisciplinary ecological governance.

(3) Advancing holistic basin governance in the middle and lower reaches

Recognizing the Yellow River Basin as an integrated system of ecological, hydrological, and sediment processes. Tailoring governance goals to regional needs, and adopting new technologies such as big data and Internet of Things for integrated monitoring and scientific management. Studying on the changes in the water–sediment relationship, vegetation restoration thresholds, and controlling sediment loss within a reasonable threshold range. Optimizing reservoir operation strategies (e.g., Sanmenxia, Xiaolangdi), balancing ecological and socioeconomic water use, and jointly advancing major ecological protection and restoration efforts across the basin.

Originally published in Chinese in *Bulletin of Chinese Academy of Sciences*, Issue 8, 2023

Authors: Fu Bojie, Liu Yansui, Cao Zhi, Wang Zhuangzhuang, Wu Xutong

Scientifically Improving Human–Nature Relationship in Desertification Areas

The second edition of the *Global Land Outlook* (*GLO2*) report, released by the United Nations Convention to Combat Desertification (UNCCD) in 2022, highlights that humanity has altered over 70% of the Earth's land from its natural state. According to the report, land degradation now affects 20–40% of the global land area, directly impacting nearly half of the world's population. This degradation poses significant challenges to achieving several United Nations Sustainable Development Goals (SDGs), including SDG 1 (No Poverty), SDG 2 (Zero Hunger), SDG 3 (Good Health and Well-being), SDG 6 (Clean Water and Sanitation), SDG 9 (Industry, Innovation, and Infrastructure), SDG 10 (Reduced Inequalities), and SDG 15 (Life on Land). The *Global Land Outlook* defines desertification as land degradation in arid, semi-arid, and dry sub-humid regions caused by climate change and human activities. The report predicts that climate change will exacerbate global desertification. SDG 15.3 explicitly calls for combating desertification, restoring degraded land and soil—including land affected by desertification, drought, and floods—and striving to achieve a land degradation–neutral world by 2030. Consequently, desertification represents a critical challenge to sustainable development in drylands, and scientifically addressing desertification is a pivotal issue for global sustainable development.

The expansion of desertification in China has been fundamentally curbed. However, the Sixth National Survey on Desertification and Sandification conducted in 2019 revealed that desertified land accounts for 26.8% of the country's total land area, while sandification land constitutes 17.6%. These findings indicate that the challenge of combating desertification in China remains severe. At the same time, as a key participant in the SDGs and an active promoter of regional sustainable development, China faces the complexity of interconnections among the SDGs. Scientifically improving the relationship between humans and nature in desertification areas to advance the synergistic achievement of SDG 15.3 and other SDGs has become a critical priority for desertification combat in China in the new era. Under the guidance of the modernization national strategy for "harmony between humanity and nature", winning the critical Three-North Shelterbelt Forest Program battle over the next decade will require a shift from a "freehand brush work" to a "claborate-style painting". This entails addressing practical challenges in the regional human-nature relationship while considering overarching issues of desertification combat and emphasizing regional characteristics. Based on theoretical insights into the coupled relationship between humans and nature, this study proposes key scientific and technological directions to improve human-natural interactions in desertification areas, providing technological support for the sustainable development of these regions.

Key Challenges to Sustainable Development in Desertification Areas

I. Increased water resource utilization

For desertification areas, the core of sustainable development lies in effectively managing the interplay among humans, water, and land, with water being the primary limiting resource and environmental condition. Water also serves as the central thread connecting the relationship between humans and land. In recent years, terrestrial water storage in northern China has shown a declining trend, with groundwater depletion identified as the major contributing factor. It is generally recognized that, excluding areas covered by snow and ice, increased human activities, particularly irrigation, are the dominant drivers behind the reduction in terrestrial water storage. For example, the oasis area in the Hexi Corridor expanded from 1.48×10^4 km^2 in 1975 to 2.14×10^4 km^2 in 2019. In the same year, the total groundwater supply in the five cities of the Hexi Corridor—Wuwei, Jinchang, Zhangye, Jiuquan, and Jiayuguan—reached 19.89×10^8 m^3. However, the over-extraction area covered 1.33×10^4 km^2, accounting for 81% of the over-extraction areas in Gansu Province, posing a significant threat to regional ecological security. Simultaneously, large-scale vegetation greening initiatives in desertification areas also impact terrestrial water storage. For instance, studies have shown that ecological restoration projects in the Mu Us Sandy Land consume approximately 16.6 ± 5.0 mm of terrestrial water storage annually. However, during the same period, extensive coal mining activities in the Mu Us Sandy Land have also altered groundwater structures, making it debatable to attribute the decline in terrestrial water storage primarily to ecological restoration projects. Overall, the increasing water demand from human activities in China's desertification areas has led to groundwater depletion, heightening the risk of terrestrial water resource shortages and posing significant challenges to achieving SDG 6.

II. Enhanced land use intensity

A common misconception in desertification combat is the simplistic belief that greener is better. In fact, in drylands, greening efforts may exceed the carrying capacity of local water resources and environments. Over the past 20 years, China and India have led global greening efforts. A closer examination of land use types reveals that in China, forests account for 42% of the greening contribution and croplands for 32%, whereas in India, croplands contribute 82% and forests only 4.4%. Cropland greening often requires irrigation and fertilizer application, significantly intensifying land use intensity. Similarly, driven by the need to ensure food security under SDG 2, regions with extensive desertification, such as Inner Mongolia and Xinjiang, have continually expanded their cultivated land and agricultural irrigation areas. For example, the agricultural irrigation area in Xinjiang and the Hexi Corridor increased rapidly from 5.81×10^4 km^2 in 2012 to 7.60×10^4 km^2 in 2021. Discussions on land use intensity in desertification areas should not be confined to the domain of territorial spatial research. The underlying scientific principles are often rooted in the water–food–ecology nexus in drylands. Relying solely on greening as an indicator of desertification combat effectiveness is questionable without disaggregating land use types, clarifying land use demands, identifying land use intensity, and considering vegetation community structures.

III. Increasing climate risks

The environment in desertification areas is highly sensitive and fragile, and global warming, coupled with an increase in extreme climate events, poses even greater challenges to the sustainable development of these regions. In particular, the higher frequency, longer duration, and greater intensity of droughts create threats to water security, food security, and ecological security in desertification areas. Evidence indicates that global warming extends the drought recovery time of ecosystems, and under more frequent drought events, ecosystems may remain in a state of incomplete recovery for prolonged periods. In the desertification areas of northern China, the drought recovery time of ecosystems is primarily determined by soil moisture. Studies have shown that soil moisture in China's semi-arid and semi-humid regions has declined, while the water demand for plant growth has increased due to the CO_2 fertilization effect. This combination has further heightened the difficulty of maintaining ecosystem stability in desertification areas. It is projected that small glaciers within Xinjiang, with areas less than 1 km^2, will completely disappear by the mid-twenty-first century. The diminishing regulatory functions of glaciers will directly impact the maintenance of regional eco-hydrological processes. Remote sensing estimates based on the critical slowing down principle indicate that vegetation resilience in China's semi-arid and semi-humid regions has declined in recent years under climate change, suggesting a reduction in ecosystem stability. Therefore, while acknowledging the significant achievements in desertification combat in China's drylands, greater emphasis is needed on the resilience, stability, and sustainability of ecosystems in drylands under future climate change.

IV. Insufficient industrial innovation drive

Desertification areas are typically located at a considerable distance from major markets, and their limited resource and environmental carrying capacity constrains the diversity of viable industries. Under the guidance of SDG 7 (Affordable and Clean Energy) and SDG 13 (Climate Action), emerging industries such as ecological carbon sinks and photovoltaic and wind energy have developed alongside desertification combat efforts in China, forming a key component of the realization of ecological product value in these regions. However, evidence suggests that, compared to southern and northeastern China, the land available for afforestation in desertification areas is extremely limited. Additionally, another study indicates that the climate benefits of afforestation in global drylands are minimal, as the reduction in surface albedo caused by afforestation leads to warming that offsets much of the cooling effect of carbon sequestration in dryland afforestation. Based on estimates of the offset relationship mentioned above, the atmospheric carbon emission reduction efficiency of photovoltaic farms in drylands is 100 times greater than that of afforestation on the same area of land. These findings highlight the inherent constraints on the development of ecological carbon sink industries in desertification areas but do not imply that photovoltaics represent a perfect solution for industrial innovation in these regions. For instance, there is a common misconception among many practitioners that photovoltaics can directly combat desertification. In reality, terms such as "photovoltaics + desertification combat" and "photovoltaics + ecology" describe parallel relationships rather than causal ones. Without corresponding sand fixation measures, the land restored under photovoltaic panels in sandification areas remains at risk of re-desertification. Current innovative industrial forms, including "photovoltaics + agriculture", "photovoltaics + forage", and "photovoltaics + traditional Chinese medicine", must still adhere to the objective principles of the water–food–ecology nexus in drylands. Gradual exploration under the principles of water availability and water resource limitations is essential, and caution must be exercised in promoting these practices on a large scale.

Scientific Understanding of Human–Nature Relationships in Desertification Areas

I. Nature's contribution to people

How to synergistically maintain or enhance nature's numerous beneficial contributions to human quality of life is a critical and challenging issue for achieving the SDGs. To address this, the Intergovernmental Science-Policy Platform on Biodiversity and Ecosystem Services (IPBES) proposed the concept framework of Nature's Contributions to People (NCP). NCP encompasses all positive contributions and negative impacts of nature on human well-being and is categorized into 10 regulating NCPs, 4 material NCPs, and 4 nonmaterial NCPs. These include aspects such as habitat maintenance, pollination, air quality regulation, climate regulation, ocean acidification regulation, freshwater regulation, water quality regulation, soil retention, hazard regulation, pest regulation, bioenergy, food, materials, medicine, learning and inspiration, experiences, identity, and options. The NCP framework represents an extension of the concept of ecosystem services, highlighting the central and pervasive role of culture in defining all connections between humans and nature. It also emphasizes the importance of indigenous and local knowledge in understanding and sustaining these relationships.

Taking the desert steppe in Northwest China as an example, its NCP is reflected not only in habitat maintenance, sand stabilization, soil retention, and food supply through pastoralism but also in the experiences and identity associated with the cultural heritage of the Northwest region. Using current ecosystem service valuation algorithms, the ecosystem service value per unit area of desert steppe is often lower than that of typical steppes in central and eastern China due to its low biomass per unit area. However, this does not diminish the importance of desert steppe. For instance, in the desertification areas of the Qinghai-Tibet Plateau, pastoralism serves as both the primary livelihood and a cultural tradition for the local Tibetan population. A survey conducted in the Qinghai-Tibet Plateau has revealed that most herders still prefer traditional free-range grazing methods and have a low subjective willingness to adopt fenced grazing restrictions. In fact, due to the intertwined nature of regulating, material, and nonmaterial contributions, it is difficult to make direct comparisons regarding the importance of different grassland and desert ecosystems. The needs of various stakeholders also vary significantly. Therefore, desertification combat should, on the basis of ecosystem service value, adopt a broader perspective when considering nature's contributions to human well-being in desertification areas.

II. Human impacts on nature

Human activities have a profound impact on the environment, many of which are widely recognized as negative, such as environmental pollution, greenhouse gas emissions, biodiversity loss, and land and soil degradation. However, since the dawn of human civilization, a mutually beneficial relationship between humans and nature has emerged, which continues to be deeply reflected in certain ethnic cultures, indigenous beliefs, and community norms. Therefore, humans and nature form a shared life community, and it is essential not only to recognize nature's contributions to humanity but also to understand people's contributions to nature. For example, one study has shown that the reciprocal relationship between humans and nature can be classified into three levels: symbolic–language–cultural, biophysical, and institutional–social–political. The first level includes empathy, moral values, the transmission of biocultural knowledge, rituals and ceremonies, oral traditions, taboos, places or species as cultural cornerstones, and the defense of natural spaces. The second level encompasses habitat enhancement, cultivation, fertilization, food provision, translocation (genetics), trait selection, restoration, and the processes of reduction–reuse–recycling. The third level includes monitoring, indigenous traditional guidelines, community-based management, environmental justice movements, international socio–ecological agreements, and the rights of nature.

China has made globally recognized achievements in ecological restoration, representing a significant example of human contributions to nature at the national scale. Among these, the Three-North Shelterbelt Forest Program stands as one of the earliest large-scale ecological projects implemented in China. It has achieved significant benefits in wind prevention, sand fixation, and carbon sequestration in desertification areas, thereby improving the relationship between humans and nature in China's drylands. Although in recent years, there have been some differing views on the Three-North Shelterbelt Forest Program, such as the argument that its climatic contributions outweigh those of human efforts and concerns over the impacts on the livelihoods of farmers and herders. Overall, if ecological engineering does not delineate restoration spaces and plant saplings and grasses, climate factors alone cannot achieve a "greening effect from nothing" in a short period. Enhancing the well-being of residents in desertification areas through high-quality development requires industrial innovation rather than relying solely on the scale and intensity of land use. Of course, the Three-North Shelterbelt Forest Program has encountered several challenges, including an unbalanced tree–shrub–grass structure and misalignment with the region's high-quality development objectives. These issues need to be addressed in the ongoing efforts to tackle desertification in the battle of Three-North Shelterbelt. This requires a more scientific understanding of the holistic, systematic, and regional nature of the human-nature relationship in desertification areas. An exemplary approach is the integration of rocky desertification combat with poverty alleviation practices in karst regions by the Chinese Academy of Sciences. They proposed alternative pastoral systems such as grass-eating livestock, the cultivation of medicinal plants, and the development of specialized fruit industries. It has formed a technological poverty alleviation industrial system that integrates vegetation-based land management with the cultivation of specialized ecological industries and the development of value-added ecological products. This approach serves as an important demonstration of how human efforts, through scientific means, can improve the relationship between humans and nature in rocky desertification areas.

III. The evolution of the human-nature relationship in desertification combat

Based on the theoretical understanding of the evolution of the human–nature coupling relationship during the ecological protection and restoration process, this study divides the desertification combat process into four stages. For a significant portion of the twentieth century, the continued development of agriculture, animal husbandry, and mining ensured food security and contributed to economic growth in many desertification areas of China. However, the early-stage extensive development led to a pattern of advancing sand dunes and retreating human settlements. The quest for food from the desert came at the cost of increased sand damage, a decline in groundwater levels, and soil salinization. These negative effects are long-lasting and persist in some regions to this day. For example, between 1999 and 2019, while precipitation and mountain runoff in the Hexi Corridor oasis significantly increased, and groundwater extraction decreased, the cumulative groundwater depth in the oasis remained relatively low, and the issue of imbalance between extraction and replenishment continues to be a serious problem.

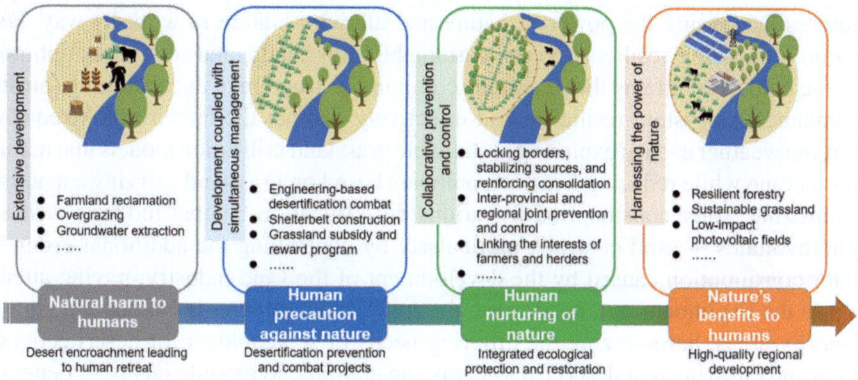

Evolution of human–nature relationship in desertification combat

From the 1980s to the first decade of the twenty-first century, with the Three-North Shelterbelt Forest Program as the leading initiative, China advanced its land greening efforts, and sand control techniques, such as grass grid projects, were gradually promoted. These efforts helped prevent further sand damage in northern China. During this period, desertification combat technologies primarily consisted of relatively independent biological, engineering, and chemical measures, with the sole aim of increasing surface vegetation cover. At the same time, due to the absence of institutional constraints such as ecological protection red lines, although sand control projects were implemented by forestry and grassland departments in specific regions, the rapid development of agriculture and mining technologies led to the continuous exploitation of ecological spaces in drylands of China. This created a pattern of simultaneous development and restoration.

Since the 18th CPC National Congress, the concept of mountains, rivers, forests as well as farmlands, lakes, grasslands, and deserts as a community of life has gradually been incorporated into desertification combat practices. For example, the governance approach of "locking borders, stabilizing sources, and reinforcing consolidation" in Qinghai, the inter-provincial and regional joint prevention and control mechanisms for desertification, such as Xinjiang–Gansu–Qinghai and Inner Mongolia–Ningxia–Shaanxi, and the sand industry development model involving enterprises, cooperatives, bases, and farmers and herders have all promoted integrated ecological protection and restoration of desertified land. While these collaborative mechanisms are commendable, it is important to recognize that desertification areas in China still face practical challenges, such as the degradation and low stability of tree plantations, the degradation of grasslands, and insufficient drought adaptability, as well as underdevelopment of the "photovoltaic + desertification combat" regional model.

In the next decade, with the high-quality development of desertification areas as the overarching goal, we should not adopt the mindset of "human conquering nature" by pursuing large-scale, high-density, high-carbon sequestration, and high-value land greening effects. Instead, we need to emphasize the characteristics, stability, adaptability, and sustainability of desert ecosystems over longer time scales. This requires harnessing the power of nature and allowing nature to work its way. For example, building resilient forests, sustainable grasslands, and low-impact photovoltaic fields; optimizing the vegetation community structure of tree and shrub forests to enhance ecosystem resilience and effectively combat the increasingly frequent extreme weather events; exploring sustainable grassland utilization models that maintain income while reducing livestock numbers based on grassland carrying capacity; minimizing water consumption in various desertification combat models, such as "photovoltaics + sand control", particularly by controlling the additional groundwater consumption caused by the development of the sand industry; paying attention to the livelihoods and local cultural traditions of residents in border and ethnic minority areas, harmonizing the differing needs of stakeholders for desert ecosystems, and ensuring that desert ecosystems can continue to provide inclusive benefits to local populations in a long-term, stable manner.

Scientifically Advancing Desertification Combat to Improve the Relationship Between Humans and Nature

I. Conducting systematic scientific research in drylands to overcome cognitive limitations

Currently, data accumulation in China's desertification areas, especially in drylands, remains relatively weak. Discussions on the underlying mechanisms of the data are mostly based on local studies, or a few region-scale indicators that are easily amenable to remote sensing inversion and model simulation. The output of systematic scientific research findings is relatively poor. China is the second largest country in terms of drylands area globally, and localized observation systems are challenging to extend from point-based data to broader regions. For example, the "Synthetic Research on Eco-hydrological Processes of the Heihe River Basin" major research program established by the National Natural Science Foundation of China has set up a systematic eco-hydrological observation network and data platform in the Heihe River Basin, providing valuable long-term time series data for basic research in drylands. However, it is clearly not feasible to directly transplant the analytical conclusions from the Heihe River Basin to other basins in China's drylands. In the past decade, the lack of major national research programs focused on drylands has led to fragmented research efforts, with many research teams working in isolation, which hinders a systematic approach to answering a series of scientific questions regarding the relationship between humans and nature in drylands. Additionally, the challenging conditions for field observations and research in drylands have resulted in many conclusions being derived through remote sensing and simulation methods. Consequently, much of the research has focused on attributes such as vegetation greenness, which is easily observed through remote sensing, while less attention has been given to properties such as soil and hydrology, for which remote sensing accuracy is relatively low. This could also mean that simulations and predictions are being carried out without sufficient local observational data to train the model parameters. Given that desertified land covers more than one quarter of China's total land area, it is recommended to establish unified standards and systems from the perspective of top-level research design, objectives and methods, with data monitoring and sharing. This involves launching a systematic scientific research program for China's drylands, promoting the creation of an integrated observation–experiment–simulation system, and establishing a multi-scale, multi-factor, multi-process collaborative observation network. This would enable a profound understanding of the interactions within the human–environment system in drylands and provide more comprehensive scientific evidence for regional desertification combat efforts.

II. Promoting nature-based solutions to prevent excessive intervention

Nature-based Solutions (NbS) aim to address diverse challenges through the efficient use of natural resources and adaptive ecosystem management, and this concept is worth promoting in desertification combat efforts. In contrast to traditional human-engineered measures, NbS focuses on actions to protect, maintain, restore, sustainably use, and manage natural or modified ecosystems in order to effectively and adaptively address social, economic, and environmental challenges, while benefiting human well-being, ecosystem services, resilience, and biodiversity. Clearly, as a primordial natural landscape, it is impossible to erase deserts from the Earth through desertification combat efforts. The eight principles of NbS can serve as a guide for effective governance rather than over-management of deserts.

(1) Effectively addressing the social challenges in desertification areas, which include not only ecological degradation but also issues such as water security, food security, climate change, and natural disasters. It is important that, while combating ecological degradation, other challenges are not exacerbated.
(2) Designing desertification combat solutions based on multiple scales, focusing on the integrity of landscape unit structures in desertification areas and their external effects, rather than solely on survival rates of seedlings at the local scale.
(3) Emphasizing net biodiversity growth and the integrity of ecosystems in desertification areas, being cautious to avoid monoculture of artificial coniferous forests.
(4) Desertification combat must also be economically feasible, and the establishment of moderate-scale ecological product value realization channels is worth advocating and fostering.
(5) Government-led desertification combat efforts should involve inclusive, transparent, and empowering governance processes, enabling stakeholders to effectively participate, and promoting increased income for farmers and herders through standardized breeding and large-scale planting.
(6) Balancing desertification control objectives with other benefits, while reasonably compensating for livelihood costs associated with grazing, cultivation,

and related activities—always with respect for indigenous cultures and local identities

(7) Adaptively managing desert ecosystems based on evidence, dynamically revising management plans for degraded forests and grasslands according to local observational data and scientific facts from desertification areas, integrating local traditional knowledge rather than rigidly copying experiences from other regions.

(8) Desertification combat measures should be sustainable and mainstreamed within the jurisdiction, encouraging pilot projects in regions with relatively sufficient funding to take the lead. For example, Ordos City, under the theme of "Desertification Combat and Green Development", is building a national sustainable development agenda innovation demonstration zone. On one hand, desertification combat measures should generally benefit future regional social development and progress. On the other hand, it also suggests that local governments in desertification areas should not separate desertification combat from their broader sustainable development planning.

III. Strengthening global research collaboration networks on human–nature interactions in drylands and enhance international influence

Desertification prevention is a global need, which requires expanding cooperation networks and strengthening scientific research on human-nature relationships in drylands worldwide. Global drylands account for about 41% of the Earth's land area, with many countries in Asia and Africa along the Belt and Road Initiative located in drylands. These countries face common challenges such as land degradation, water crises, and regional development stagnation. However, many developing countries have relatively limited scientific research capabilities in the field of desertification prevention. In 2017, the Chinese Academy of Sciences initiated the "Global Dryland Ecosystem International Science Program" (Global-DEP) to focus on social–ecological systems in drylands. The main research areas of the program are divided into four parts: dynamic changes and driving mechanisms of dryland social–ecological systems, changes in the structure and function of dryland social-ecological systems, ecosystem services and human well-being in dryland, and sustainable livelihoods in dryland. To support this initiative, the Global Land Programme (GLP) established the "Global Dryland Social–Ecological Systems" working group. Under this project, scholars both domestically and internationally have collaborated to publish the research monograph *Dryland Social-Ecological Systems in Changing Environments.*

Recommendations for enhancing the international influence of research in desertification prevention are proposed from both "bringing in" and "going out" perspectives: ① Rooted in key regions: large-scale scientific projects with stable funding should be implemented in key regions, focusing on specific shared interests such as ecological restoration, climate adaptation, water security, food security, and disaster prevention. Through bilateral and multilateral cooperation in the form of targeted technical services, these efforts can provide new insights, technologies, equipment, and solutions for achieving harmonious coexistence between humans and nature in desertification areas of China. ② Broaden international perspectives: further integrating international expertise and think tanks in the field of desertification prevention, focusing on key desertification areas along the Belt and Road such as Mongolia, Central Asia, West Asia, and the Sahel. Collaborating with Chinese expert teams, joint research efforts should be made to contribute to global desertification combat, enhancing China's international image and scientific influence.

Originally published in Chinese in *Bulletin of Chinese Academy of Sciences*, Issue 12, 2024

Authors: Fu Bojie, Liu Yanxu, Wang Shuai, Song Junfeng, Li Zejiang

Scientifically Conducting Territorial Ecological Restoration

The ecological restoration of territorial space, through the protection, restoration, and reconstruction of ecosystems, provides a specific operational guide for improving ecosystem quality and stability and safeguarding the boundary of natural ecological security. It is also a direct embodiment of the concept that "lucid waters and lush mountains are invaluable assets". Considering China's vast territorial area, the diversity of ecosystem types, and the disparities in levels of social development, a deep understanding of the human–nature relationship across different regions is a scientific prerequisite for supporting territorial spatial ecological restoration and promoting harmonious coexistence between humans and nature. In light of the overall requirements for high-quality development during the 14th Five-Year Plan period, it is proposed to deepen the understanding of the natural regional system from the following four aspects, in order to lay a solid scientific foundation for the comprehensive implementation of territorial spatial ecological restoration.

First, it is recommended to start from the integrity of natural regions to identify key areas for territorial spatial ecological restoration and safeguard the boundary of natural ecological security. Natural ecological security boundaries often do not align with administrative boundaries, especially in the case of large natural regional units that function as major ecological security barriers across provinces. This requires territorial spatial ecological restoration to have a clear understanding of the national-scale ecological security strategic layout. During the 13th Five-Year Plan period, China's national main functional zone planning proposed an ecological security strategic layout centered on the "two barriers and three belts": the Qinghai-Tibet Plateau ecological barrier, the Loess Plateau–Sichuan–Yunnan ecological barrier, the northeastern forest belt, the northern sand prevention belt, and the southern hilly and mountainous belt, forming an overall ecological outline for green development. In the 14th Five-Year Plan period, China's ecological security strategic layout should shift from "freehand brush work" to a "claborate-style painting" by refining the national ecological security pattern through continued implementation of the main functional zone strategy. This will help establish the basic framework of China's natural ecological security boundary and provide overall spatial guidance for ecological restoration.

For example, the Loess Plateau–Sichuan–Yunnan ecological barrier is not a single unit in terms of physical geography and ecological zoning. The Qinling Mountains, which traverse its central part, mark the dividing line between northern and southern China. The Qinba Mountains possess a range of significant ecological functions such as maintaining biodiversity, conserving water resources, and regulating climate, and they deserve to be designated as a national ecological security barrier, holding a position of equal importance to the Qinghai–Tibet Plateau and the Loess Plateau. Another example is the coastline, which serves as the "lifeline" of marine economic development. In recent years, there has been a sharp reduction in natural shorelines. According to China's first coastal law, the national natural shoreline retention rate was required to be no less than 35% by 2020. During the 14th Five-Year Plan period, the inclusion of coastal zone protection in the national ecological security layout will provide stronger support for the comprehensive ecological protection and restoration of the coastal zone from the perspective of integrated land–sea coordination.

Second, it is recommended to conduct in-depth research on the interactions and effects among various ecological elements such as mountains, rivers, forests, farmlands, lakes, grasslands, and deserts, determine ecological protection and restoration approaches based on local conditions, and improve the quality and stability of ecosystems. Ecological protection and restoration approaches can be categorized into the conservation of specific species or habitats, protection of regional ecology, restoration of damaged ecosystems, and reconstruction of ecosystems that have lost their self-recovery capacity. In humid and semi-humid areas, when the degree of ecosystem degradation is relatively low, natural recovery can be achieved more rapidly by protecting ecological space and preventing human disturbance. However, in arid and semi-arid regions, where natural ecological succession proceeds slowly, it is often necessary to adopt artificial-assisted restoration or ecological reconstruction by planting shrubs, trees, or herbaceous plants. It is worth noting, however, that the various ecological components—mountains, rivers, forests, farmlands, lakes, grasslands, and deserts—are interrelated. Increases in vegetation in arid areas often alter the water cycle, while the threshold constraints of soil moisture and groundwater on ecosystem stability in these regions remain uncertain. Therefore, it is essential to recognize that greener terrestrial ecosystems are not necessarily better. Artificial ecological restoration and reconstruction should be grounded in natural geographical laws and adhere to the principles of prioritizing quality and stability. It is necessary to scientifically assess the planting suitability of regional zonal vegetation and, based on a clear understanding of the interactions among ecological elements and their resource–environment effects, determine restoration approaches and spatial configurations of mountain–river–forest–farmland–lake–grassland–desert landscapes suited to local natural regional conditions.

Third, it is recommended to further integrate ecological restoration into economic, social, and cultural development, support the optimization of territorial spatial patterns and the improvement of people's livelihood and well-being, and promote a comprehensive green transformation of economic and social development. Territorial spatial ecological restoration is not confined to natural ecosystems alone, but involves a wide range of spatial elements including ecological areas, urban spaces, farmlands, and industrial and mining zones. Under the concept that "lucid waters and lush mountains are invaluable assets", a key mission of such restoration is to enhance productive capacity and improve regional livelihoods and well-being. The central subject of ecological security is people, and its goal is to maintain and improve human well-being while preserving ecosystem functions. Therefore, promoting people's livelihoods and well-being further clarifies the goals and benefits of territorial spatial ecological restoration from the perspective of societal demand. During the 13th Five-Year Plan period, China's territorial spatial ecological restoration, through the implementation of mountain–river–forest–farmland–lake–grass ecological protection and restoration projects, laid a solid ecological foundation for green development in certain regions. In the 14th Five-Year Plan period, on the basis of continued project implementation, it is necessary to strengthen the core role of territorial spatial ecological restoration in production and daily life. This includes the enhancement of farmland productivity and ecological functions in the context of rural ecological revitalization, the improvement of urban ecological quality under the goal of new urbanization, and the development of green mines under the trend of energy transition. From the perspective of optimizing territorial spatial patterns, efforts should be made to fully implement the integration of ecological civilization into political, cultural, economic, and social development.

Fourth, it is suggested to establish a unified national ecosystem and natural resource monitoring system to facilitate data integration, analysis, and sharing. Scientific data serve as the foundation for understanding natural laws and bridging the gap between knowledge and decision-making. Over the past 30 years, ecosystem monitoring networks globally, especially in developed countries, have been continuously improved, resulting in systems such as the International Long-Term Ecological Research Network (ILTER), the Terrestrial Ecosystem Monitoring System (TEMS), the global FLUXNET network, the National Ecological Observatory Network (NEON) in the United States, and the Environmental Change Network (ECN) in the United Kingdom. China's Ecological Research Network (CERN) and the National Ecosystem Observation and Research Network (CNERN) have also had a significant impact within the domestic scientific community. In 2020, the Ministry of Natural Resources released the Overall Plan for the Construction of the Natural Resources Survey and Monitoring System, which systematically reconstructed the indicator system for natural resource surveys, introduced a three-dimensional classification of natural resource spatial levels, and proposed a comprehensive special survey of natural resources. The overall design of the plan is considered advanced. Given that ecosystem monitoring often requires long-term time series datasets, it is difficult for natural resource surveys to achieve full coverage of all natural regional systems in China within a short time using site-based observation methods. During the 14th Five-Year Plan period, it is necessary to integrate more effectively monitoring station data from subordinate departments of the Ministry of Natural Resources, Ministry of Ecology and Environment, Ministry of Water Resources, Ministry of Education, China Meteorological Administration, and Chinese Academy of Sciences. It is also essential to optimize the analytical indicators used at different spatial scales for ecosystem degradation assessment, ecological restoration suitability evaluation, and ecological restoration effectiveness assessment. Mechanisms for interdepartmental data submission, sharing schemes, and management methods should be formulated. This will provide indicator-based, comprehensive, and scale-matched scientific data support for gaining an in-depth understanding of natural regional system characteristics and for carrying out territorial spatial ecological restoration in a scientifically informed manner.

Originally published in Chinese in *Natural Resources Information*, November 2020

Authors: Fu Bojie, Liu Yanxu, Zhao Wenwu, Wang Shuai

Using Ecological Science and Technology to Create a Beautiful Homeland

The construction of ecological civilization aims at achieving harmonious development between humanity and nature. The ecological concept 1.0 focused on the coordination of elements. People became aware of the limitations of resource and environmental carrying capacity under high-intensity human activities, and began exploring how to coordinate the relationship between humans and nature through resource conservation and environmental protection. Based on this concept, targeted measures such as returning farmland to forests, establishing nature reserves, and controlling pollution were implemented. As the ecological concept evolved into 2.0, emphasis shifted to systemic governance. People came to understand that the constraints imposed by resource and environmental elements on human society are not isolated, and that systemic problems must be addressed through systemic solutions. Projects for the ecological protection and restoration of mountains, rivers, forests, farmlands, lakes, and grasslands, the construction of the nature reserve system, and campaigns for source pollution control all reflect the holistic and systemic thinking of ecological civilization construction. The ecological concept 3.0 advocates for deep harmony between humanity and nature. Society will enter a path of green development, where development and conservation shift from a trade-off relationship to one of mutual synergy and co-existence. Ecological and environmental benefits will become a driving force for socioeconomic growth, and lucid waters and lush mountains give rise to gold and silver mountains.

The ongoing advancement of ecological civilization cannot be achieved without the power of science and technology. Whether it involves optimizing the territorial spatial development pattern, developing energy-saving and low-carbon industries, building a water-saving society, improving the disaster prevention and reduction system, or establishing a system of protected natural areas, science and technology play an important role in all these efforts. Technological innovation and application have been deeply integrated into all aspects of environmental governance, ecological protection, and green development. For example, the promotion of new energy sources such as solar and wind power, the implementation of basin-wide comprehensive management to address lake pollution, and the scientific execution of ecosystem protection and restoration projects all reflect this integration. It can be said that the path toward green, low-carbon, and high-quality development is also a path through which technological innovation contributes to ecological protection.

Smart Environmental Protection: The Construction of Ecological Civilization Is Closely Related to the Application of Science and Technology

Promoting the development of new energy is essential. It is well known that the use of fossil fuels emits large amounts of carbon dioxide and other greenhouse gases, leading to global warming, more frequent extreme weather, and increasing disaster events. The development and utilization of new energy sources such as solar and wind power are key priorities for future energy development. Over the past decade, China has developed large-scale photovoltaic and wind power projects in arid regions, yielding considerable economic, social, and ecological benefits. Simulation studies have shown that the large-scale deployment of wind and solar power facilities in arid desert regions can promote regional precipitation and vegetation recovery. These developments have no significant negative impact on regional climate and are conducive to improving regional climate and ecological conditions. These research findings provide scientific support for the green development of arid regions from the perspective of the energy–water–food nexus. The promotion of new energy and the enhancement of ecosystem carbon sinks have become important measures for actively addressing climate change and achieving carbon neutrality.

Improving environmental quality is essential. Non-point source pollution caused by the intensive use of pesticides and chemical fertilizers, such as water and soil pollution, is more difficult to control and manage than point source pollution represented by industrial emissions. In particular, the deterioration of river and lake water quality caused by non-point source pollution causes symptoms in the water bodies, but its root causes lie within the entire basin. Wuliangsuhai, located at the northernmost end of the Yellow River, is the eighth largest freshwater lake in China. At present, the local government is promoting a series of environmental and ecological engineering technologies, including comprehensive treatment of canal system aquatic plants and wetlands, reuse of reclaimed water, resource-based utilization of aquatic vegetation, bottom mud disposal, precision fertilization in irrigation districts, green pest control technologies, integrated water and fertilizer management, monitoring and disposal of plastic film residues, and resource utilization of livestock and poultry manure. Through farmland measures of "controlling fertilizer, pesticide, water, and plastic film" along with wetland restoration near the lake, the water quality of Wuliangsuhai has improved from below Grade V to overall Grade V, and in some areas has reached the Grade IV standard. This demonstrates that a pollution control and management technology system led by integrated basin management can effectively address prominent regional water environment problems.

Scientific planning and implementation are essential for ecological restoration. Vegetation restoration projects, such as those involving the conversion of farmland to forests and grasslands, have increased vegetation coverage and significantly improved various ecological functions. However, research and practice in arid and semi-arid regions demonstrates that, if the water resource constraints of large-scale vegetation restoration are neglected, it may lead to a significant decline in basin runoff and soil moisture, resulting in resource waste such as "planting trees year after year without seeing trees grow". It can also alter the water cycle and affect water availability for downstream production and domestic use. Only through scientific planning that comprehensively considers regional water production, consumption, and demand can a vegetation structure and restoration model suited to local environmental conditions be established. In the scientific implementation of ecological restoration, the Huanjiang Karst Ecological Station of the Chinese Academy of Sciences in Guangxi has integrated ecological restoration with poverty alleviation efforts. In areas with severe rocky desertification, ecological migration has been implemented. In the areas of origin, alternative grass-fed livestock industries combining crop and livestock production have been developed. In the resettlement areas, the advantages of local water and soil resources have been utilized to develop high-value ecological orchards and economic forests. This has formed a synergistic model integrating rocky desertification control with ecological industry-based poverty alleviation. It serves as a typical national model for both rocky desertification control and targeted poverty alleviation, and has been recognized as one of the best global cases of poverty reduction through ecological restoration.

In recent years, technical challenges in energy conservation and efficiency improvement, circular economy, low-carbon transportation, and green buildings have been continuously overcome. Some technological breakthroughs have drawn public attention, such as the precise prevention and control of air pollution. Researchers have made significant progress in understanding the causes of atmospheric haze and in developing control technologies, and pollution forecasting and early warning capabilities have also improved considerably. The rapid development of information technology has brought unprecedented opportunities for environmental protection and ecological governance. Technologies such as big data, artificial intelligence, and geographic information systems have helped people collect information more efficiently, monitor ecological and environmental conditions more accurately, and make more informed decisions. A good example is the establishment of a satellite communication station in the Hoh Xil Nature Reserve, which has enabled internet access, satellite television, and remote video communication. With the help of cutting-edge technologies, ecological civilization construction has entered the era of "smart environmental protection".

Sponge City: Urban Ecological Construction Should Be Tailored to Local Conditions and Innovate Governance

When people talk about ecology, they often think of mountains, rivers, and other natural ecosystems. In fact, cities—being the primary living spaces for humans—are crucial areas for the construction of ecological civilization. People often equate ecological cities with garden cities or green cities, but these are only partial representations of what an ecological city entails. Urban ecological construction is not merely for beautification purposes; it must also take ecological functions into account. Ecology should be integrated into the city, rather than used as decoration. Urban ecological construction has become an important vehicle for putting scientific concepts into practice and applying technological innovations. The following section illustrates this using the "sponge city" as an example.

A sponge city is an urban rainwater and flood management concept under the framework of ecological civilization. It refers to a city that, like a sponge, possesses strong resilience in adapting to environmental changes and responding to natural disasters. During rainfall, it can absorb, store, infiltrate, and purify water, and when needed, release and utilize the stored water. Compared with traditional cities, where heavy rainfall tends to accumulate on impervious surfaces and cause urban waterlogging, sponge cities achieve water retention and purification through the construction of ecological systems such as vegetated swales, permeable bricks, rain gardens, sunken green spaces, urban wetlands, and urban forests. If, in the planning and development of new urban areas, we prioritize the preservation of natural ecological elements such as green spaces and water systems, and then design production and living spaces around these ecological elements, we can effectively utilize these natural ecological corridors to fulfill their functions in ecological conservation, drainage, and flood control.

Jinhua City in Zhejiang Province is one of the earliest cities in China to adopt the sponge city construction. The ecological concept and technology of "managing water with water" have been applied and promoted here. By utilizing the capacity of buildings, roads, green spaces, and water systems to absorb, store, infiltrate, and slowly release rainwater, the city gains ecological resilience and behaves like a sponge. An ecological environment characterized by "no waterlogging in light rain, no urban flooding in heavy rain, no black and odorous water bodies, and mitigated urban heat islands" has benefited the local residents.

Integrating ecology into the city means allowing the elements of the ecosystem to perform their respective ecological functions, building an urban ecological network, and enhancing ecosystem connectivity. Ecosystems possess hierarchical and network characteristics. Only when a network is formed can the system be stable, have greater and stronger carrying capacity, and be organically integrated with the human living environment. The Guangdong 10,000 Li Greenway Master Plan, released in 2020, takes the rivers, lakes, and their banks in Guangdong as the framework and coordinates ecological, safety, cultural, landscape, leisure, and economic functions. It aims to build an urban agglomeration ecological network with the water system as the carrier, becoming an innovative design case for the construction of ecological civilization in cities and providing important planning references for the integration of mountains, waters, forests, fields, lakes, grasses, and cities.

Urban ecology has its particularities. In applying technology, it is necessary to fully integrate the characteristics of urban production and daily life, adapt to local conditions, and promote innovative governance. Urban living and production activities on the one hand require large amounts of energy, and on the other hand lead to the proliferation of waste. Using the heat generated from waste incineration for electricity generation not only reduces the need for fossil fuel extraction but also prevents soil and water pollution caused by landfill. Waste-to-energy can be achieved in two main ways. First, high-temperature incineration is used to convert thermal energy into high-temperature steam, which drives turbines for electricity generation. Second, organic waste is fermented to produce biogas, which is then burned to convert heat into high-temperature steam. Given the differences in power generation methods and the fact that some types of waste can produce harmful gases when directly incinerated, promoting waste sorting is necessary. The waste sorting systems currently implemented in cities such as Beijing provide an important foundation for the comprehensive utilization of waste and the application of waste-to-energy technology, and constitute a concrete measure in the construction of ecological civilization.

China's construction of ecological civilization will contribute Chinese wisdom, Chinese solutions, and Chinese experience to the global achievement of sustainable development goals. We look forward to more scientific understanding and advanced technologies being applied to the construction of ecological civilization, enabling green development through technological innovation, creating a more beautiful landscape homeland and building a better life for humanity.

Originally published in Chinese in *People's Daily*, February 2, 2021

Integrating Ecology into Cities

Nowadays, many cities simply equate eco-cities with urban beautification—planting flowers and grass—but this is only one aspect of ecology, not its entirety. Building an eco-city requires that the elements of the ecosystem perform their respective ecological functions. It's not just about looking good—more importantly, it's about being functional. Ecology should be integrated into the city, not merely used to decorate it.

What functions do ecosystems serve? From a scientific perspective, ecosystems provide various services to humanity and form the foundation for sustainable human survival and development. These services can be classified into provisioning services, regulating services, cultural services, and supporting services. Provisioning services refer to products that humans obtain directly from ecosystems, such as food, timber, and fuel. Regulating services refer to the benefits people gain from the regulation of ecosystem processes, including water retention, flood regulation, and climate regulation. Cultural services are those provided through recreation, aesthetic enjoyment, and similar experiences. Supporting services are the underlying processes that make all other ecosystem services possible.

To integrate ecology into cities, we need to utilize the functions of ecological infrastructure such as forests, grasslands, wetlands, and rivers. Take sponge cities as an example: by leveraging the rainwater absorption, retention, infiltration, and slow-release functions of buildings, roads, green spaces, and water systems, cities can become sponge-like and ecologically resilient. This can effectively mitigate the urban flooding caused by intense rainfall accumulating on hardened surfaces in traditional cities and reduce the urban heat island effect.

How can we integrate ecology into cities? Ecological planning should run through the entire process of urban development. Urban development should be based on an analysis of the resource and environmental carrying capacity of a region or watershed, treating the city as an integrated ecosystem. It is necessary to strengthen the construction of urban ecological infrastructure, build ecological networks, optimize the city's ecological structure and functions, and couple the city's regulating and cultural services. This will create a "community of mountains, rivers, forests, farmlands, lakes, grasslands, and cities", and build ecological landscapes where humans and nature coexist in harmony. Many of today's severe urban floods from small streams are the result of damaged ecological infrastructure. The original urban areas had natural rivers, creeks, wetlands, green spaces, and forests. These natural ecological corridors have functions of environmental protection and drainage/flood prevention. If we transform them into standardized, leveled areas or completely artificial infrastructure, it becomes difficult for them to fully perform their ecological functions. Therefore, in the process of urban planning and development, we need to prioritize preserving the "backbone" of natural ecological infrastructure such as green spaces and water bodies. Then, based on these ecological elements, we can lay out the spaces for production and daily life, forming a rational structure and pattern.

How to Understand "Lucid Waters and Lush Mountains Are Invaluable Assets"

"Lucid waters and lush mountains are invaluable assets"—this profound and philosophical statement is now widely known and deeply rooted in people's hearts. It is a succinct summary of humanity's understanding of the value of nature. So, what exactly are "lucid waters and lush mountains"? From an ecological perspective, they refer to healthy and stable ecosystems such as forests, grasslands, wetlands, shrublands, and farmlands, which are the foundations for human survival and development.

Why do we say "lucid waters and lush mountains are invaluable assets"? Ecosystems can provide a wide range of services for humanity. They have significant ecological and economic value and can generate economic benefits to improve human well-being. These services can be categorized into four types: provisioning, regulating, cultural, and supporting. Provisioning services provide material goods essential for human life, such as fresh water, food, fuel, and timber. Regulating services create and maintain the environmental conditions necessary for human survival, including climate regulation, water retention, soil conservation, flood control, disease regulation, and pollution reduction. Cultural services refer to the benefits ecosystems provide through aesthetic experiences, tourism, recreation, mental health promotion, and more. Supporting services include the fundamental ecosystem processes, such as soil formation and nutrient cycling.

The value of the services provided by ecosystems is immense. The earliest proposal for calculating the value of ecosystem services came from Professor Robert Costanza of the University of Maryland. In 1997, he published an article in the journal *Nature* estimating that the world's ecosystems provide services worth 33 trillion US dollars annually—1.8 times the global Gross National Product (GNP) at that time. This indicates that the value of ecosystem services far exceeds people's intuitive understanding. If ecosystems are destroyed, humanity loses the foundation for long-term survival. We must firmly establish and put into practice the idea that lucid waters and lush mountains are invaluable assets, and plan development from the high ground of harmonious coexistence between humanity and nature. Establishing and improving mechanisms for realizing the value of ecological products is a key pathway to implementing this concept and is of great significance in promoting a comprehensive green transformation of socioeconomic development.

Afforestation: The More, the Better?

Over the past 20 years, China has made remarkable achievements in ecological construction. Both forest area and stock volume have continuously increased, making China the country with the largest growth in forest resources globally. According to the results of the Eighth National Forest Resources Inventory, China's forest area has reached 208 million hectares, with a forest stock volume of 15.137 billion cubic meters and a forest coverage rate of 21.63%.

Is planting trees always a case of "the more, the better"? The general public tends to believe that planting trees is always beneficial. However, ecological construction must conform to the local natural environmental conditions and characteristics. Greener is not always better, and planting more trees is not always beneficial. In arid regions where trees are naturally scarce, afforestation efforts often face two problematic paths: one is to divert water for irrigation, which may strain already limited water resources; the other is to plant trees without sufficient water support, risking depletion of groundwater and soil moisture. Take the Loess Plateau as an example. In some areas, unreasonable afforestation has consumed a large amount of soil moisture, expanded the area of dry soil layers, and triggered phenomena such as "little old man trees" or even tree death, resulting in the degradation of vegetation communities. Therefore, we proposed that management of the Loess Plateau should follow the principle of converting farmland back into forests or grasslands—reforesting where conditions are suitable for trees, and planting grass where trees are not feasible. This is because grasses have relatively shallow roots and limited water demand.

Nowadays, the concept of ecological restoration has evolved further, advocating for nature-based restoration and creating ecosystems that are as close to natural conditions as possible. There are various approaches to ecological protection and restoration—not just tree planting and a series of engineering-based restoration efforts. On one hand, we protect ecosystems to allow for natural recovery, enabling them to adapt to natural conditions and provide long-term, stable ecological functions. On the other hand, artificial assistance or ecological reconstruction may also be necessary.

How should we choose the ecological restoration strategy? This depends on the degree of ecosystem degradation. In areas where degradation is relatively mild and natural restoration is possible, passive protection measures like enclosure can be adopted to reduce human disturbance and allow the ecosystem to recover naturally, so that it can adapt to local climate, soil, and terrain conditions. In more severely degraded areas, artificial assistance is needed. For example, in the early stages, some grasses and saplings can be planted, and then, after the ecosystem reaches a certain level of recovery, natural restoration can follow. By aligning the degree of degradation with appropriate recovery mechanisms and policies, we can gradually enter a new phase of ecological restoration characterized by coordination between humans and the environment, and harmony between the natural world and human activities.

Building "Four-Season Gardens" with an International Perspective

Not long ago, General Secretary Xi Jinping delivered an important speech at the meeting marking the 30th anniversary of the establishment of Hainan Province and the Special Economic Zone, proposing that Hainan Province should build a national ecological civilization experimental zone to explore new experiences for advancing ecological civilization construction nationwide.

Hainan's unique environment is a prominent advantage in ecological civilization construction and also the foundation for its sustainable development. However, from the perspective of the ecosystem itself, Hainan is an island-type ecological system, which is generally fragile and difficult to restore once damaged. Therefore, ecological civilization construction in Hainan must follow an integrated planning approach for mountains, waters, forests, farmlands, grasslands, seas, and cities, with systematic protection at its core, in order to build a rational ecological spatial layout of the territory.

Since 2016, China has established national ecological civilization experimental zones in Fujian, Jiangxi, and Guizhou. These zones have achieved preliminary results in promoting and facilitating ecological civilization construction. For example, Fujian has focused on a series of reform measures, including the ecological and environmental protection target responsibility system, whole-basin ecological compensation, comprehensive management of small watersheds, ecological judicial protection, natural resources departure audits for party and government leaders, environmental rights trading, and green finance, all of which have yielded significant results.

Jiangxi has established and improved red lines for ecological protection, water resources, and land resources, refined the property rights system for natural resources and the spatial control system, comprehensively implemented the river chief system and whole-basin ecological compensation, and improved assessment and evaluation mechanisms such as target evaluation for ecological civilization construction and accountability for ecological environmental damage.

Guizhou has carried out extensive exploration on issues such as the property rights system of natural resource assets, the territorial spatial development and protection system, the paid use of resources and the ecological compensation system, the market system for environmental governance and ecological protection, and performance evaluation and institutional accountability in ecological civilization construction.

In other provinces, provincial-level spatial planning pilots and natural ecological space use control pilots have also yielded corresponding experience. The national park system pilot, which has been implemented for nearly three years, is also being actively advanced. Overall, although the actual conditions vary across provinces, some of these practices provide valuable references for the construction of ecological civilization in Hainan.

It is worth noting that in 2010, Hainan initiated the construction of the International Tourism Island. The development goal of the International Tourism Island is to achieve sustainable socioeconomic development in Hainan by leveraging the advantages of its natural environment, making it a true embodiment and practical carrier of ecological civilization construction. This means that ecological civilization construction in Hainan must be guided by an international perspective.

For example, Hainan has a typical tropical ecosystem and a unique island ecosystem, so it can draw on domestic and international experience in national park construction and protect biodiversity and unique ecosystem types by establishing national parks such as tropical rainforests, seasonal rainforests, and tropical island ecosystems, in order to further enhance the service capacity of ecosystems. At the same time, by further improving the natural ecological spatial structure and building ecological networks of coastal zones, forests, wetlands, and other elements across the whole region, the regulation, cultural, and tourism services of ecosystems can be enhanced. Hainan Island' s excellent environment and suitable industrial structure are key pathways to achieving the goals of Hainan' s ecological civilization experimental zone; therefore, it is necessary to vigorously develop ecological industries, promote ecological culture, and improve ecological quality.

Implementing and advocating a circular economy model, with an emphasis on the efficient utilization of resources and waste, has become an important pathway for advancing ecological civilization construction in the process of modernization. Therefore, the reform of Hainan's ecological civilization system needs to focus on the institutional framework, territorial space development, ecological protection, and green production and life styles.

At the same time, Hainan Island has unique geographical and ecological advantages, which are key factors that distinguish its ecological civilization construction from that of other provinces. Therefore, Hainan's national ecological civilization experimental zone needs to implement the strictest ecological and environmental protection systems and establish a modern regulatory framework for ecological and resource protection. It is necessary to strictly protect the marine environment and establish a well-functioning regional coordination mechanism for ecosystem protection, restoration, and pollution prevention that integrates land and sea management. At the same time, a feasible ecological compensation mechanism should be established, and pilot projects for green GDP accounting should be promoted. Extensive ecological civilization publicity, education, and scientific outreach should be carried out to continuously strengthen citizens' ecological civilization concepts and environmental protection awareness. The role of communities, environmental protection organizations, and other social groups should be fully leveraged to foster a favorable atmosphere of broad societal attention and participation.

Originally published in Chinese in *China Science Daily*, May 14, 2018

Challenges and Recommendations for Achieving the Sustainable Development Goals

The 17 United Nations (UN) Sustainable Development Goals (SDGs) serve as a global vision and action framework for a more just, inclusive, and sustainable future. Since the adoption of "Transforming Our World: The 2030 Agenda for Sustainable Development" in 2015, the 17 SDGs have guided countries worldwide toward shared prosperity and sustainability from 2016 to 2030. They address critical issues such as poverty, inequality, climate change, environmental degradation, and deficiencies in health and education, reflecting international consensus on collaborative and inclusive social change. However, realizing this vision demands coordinated efforts from governments, businesses, organizations, and individuals at multiple levels.

China, as a major developing country, has not only participated in shaping the SDGs but has also actively promoted them. Despite significant progress in poverty reduction, environmental governance, technological innovation, and health and education, China still confronts challenges like urban–rural disparities, resource inequality, and pollution. With less than six years until 2030, it is alarming that while some progress has been made, many goals have stalled or regressed, potentially preventing the global achievement of most SDGs by 2030. Thus, identifying key challenges and seeking solutions is of utmost importance.

SDGs Progress Globally and in China

According to the 2023 UN *Sustainable Development Goals Report* and the 2023 *Sustainable Development Report* by the Sustainable Development Solutions Network (SDSN), global SDG progress has been limited. Positive developments include a decline in extreme poverty and child mortality, advancements in gender equality, increased electricity access in the poorest nations, growing renewable energy adoption, and a return of global unemployment rates to pre-2008 financial crisis levels. However, progress on most goals remains fragile and slow. The COVID-19 pandemic, regional conflicts, and climate change have severely impeded SDG implementation. Data shows over 30% of goals have stagnated or regressed below 2015 levels. If current trends continue, 575 million people will still live in extreme poverty by 2030, hunger levels may revert to 2005 levels, food prices will rise in many countries, renewable energy will remain a small part of energy supply, and 660 million people will lack electricity.

Since 2016, China has integrated the 2030 Agenda with its medium- to long- term development strategies, such as the 13th and 14th Five-Year Plans and the Vision 2035 of the PRC. By 2023, China had released three progress reports, participated twice in voluntary national reviews, and shared implementation experiences. Key achievements include eradicating extreme poverty by 2020, with 98.99 million rural residents lifted out of poverty, and maintaining steady economic growth, reaching a 2022 GDP of 121.02 trillion yuan. China has also shown significant environmental progress, with improved air quality, better surface water quality, soil pollution risk control, and a 15.5% decrease in energy consumption per unit of GDP from 2015 to 2022.

However, China's SDG Index, while rising from 69.42 in 2016 to 72.01 in 2023 (indicating 72% goal completion), ranks 63rd among 166 assessed countries. Among the 17 SDGs, only SDG 1 (No Poverty) and SDG 4 (Quality Education) are near completion. Twelve SDGs show slow or no progress, and SDG 15 (Life on Land) is declining, highlighting ongoing challenges.

Major Global Challenges to SDG Achievement

Different countries face unique challenges in achieving the SDGs, reflecting environmental, economic, cultural, and political differences. However, common challenges can be identified:

I. Data gaps and monitoring limitations

Since 2015, less than half of 193 countries or regions have reported internationally comparable data. Resource-constrained and technologically underdeveloped nations struggle to collect sufficient and reliable data for SDG monitoring, hindering problem diagnosis. Data timeliness is also a concern, with less than 30% of the latest data from

2022 and 2023, and over half from 2020 and 2021. Although China has improved data collection, it still faces issues of data quality and regional availability. For example, nine SDGs at the provincial level in China have indicator numbers below 50% of the official SDG indicators. These data gaps significantly affect accurate SDG implementation assessment and monitoring.

II. Complexity of SDG interactions

The 17 SDGs span economic, social, and environmental domains, with both synergies and trade-offs between goals. For instance, if excessive fertilizers and pesticides are used, poverty (SDG 1) and hunger (SDG 2) reduction through increased agricultural production may negatively impact water quality (SDG 6) and terrestrial ecosystems (SDG 15). Similarly, industrialization (SDG 9) can drive economic growth (SDG 8) but may exacerbate climate change (SDG 13) without proper management. In China, this complexity is evident as the country needs to balance clean energy adoption (SDG 7) with industrial upgrading (SDG 9), ensuring economic growth does not compromise environmental quality (SDG 11 and SDG 12).

III. Regional differences and negative externalities

Differences in environmental conditions, resources, economic development, and public services lead to varying national and regional development priorities. Developing countries may focus on infrastructure (SDG 9) and poverty (SDG 1) and hunger (SDG 2) eradication, while developed nations prioritize climate change (SDG 13) and environmental protection (SDG 12 and SDG 15). Within China, coastal regions may emphasize innovation and sustainable industrialization (SDG 9), whereas inland areas focus on poverty (SDG 1), education (SDG 4), and economic development (SDG 8). Economic activities like industrial production or agricultural expansion in one region can cause transboundary pollution and resource depletion, affecting other countries and global well-being. Promoting regional cooperation through institutional and economic incentives is crucial for SDG achievement.

IV. Weak science–policy–society nexus

Insufficient connections between science, policy, and society pose a global challenge to SDG implementation. Despite efforts to establish SDG coordination institutions since 2015, their effectiveness remains unverified. Scientific research often faces time lags and comprehension barriers in policy-making, limiting its application. Public awareness and participation in the SDGs are also critical but hindered by inadequate information communication mechanisms. Strengthening science–policy dialogue and enhancing public understanding and participation are key challenges to address.

V. Uncertainty arising from global changes

Global challenges like climate change, environmental degradation, resource depletion, and political and economic shifts increase uncertainty for SDG implementation. Climate change can worsen poverty (SDG 1), food security (SDG 2), and water supply (SDG 6), while ecosystem degradation threatens marine and terrestrial conservation goals (SDG 14 and SDG 15). Political tensions, potential relationship deterioration, and rising economic decoupling further complicate SDG achievement. Countries need to balance addressing global changes with progressing SDGs. Enhancing international cooperation and adopting inclusive policies are vital for mitigating these uncertainties.

Recommendations for Advancing SDGs in China

To address slow or regressing global SDG progress, scientists propose six key transformations: ① enhancing human well-being and capabilities; ② shifting toward sustainable and equitable economic models; ③ establishing sustainable food systems and healthy nutrition patterns; ④ decarbonizing energy and improving clean energy access; ⑤ promoting sustainable urban and peri-urban development; ⑥ securing global environmental commons. Considering diverse national and regional resource endowments, development needs, and governance systems, a "classification–coordination–collaboration" approach is gaining attention, with economic, technological, and cultural collaboration seen as essential for these transformations.

In view of the urgency of accelerating the achievement of the SDGs, the following four-aspect action recommendations are put forward by comprehensively considering the concerns of the academic community, governments and the general public, with the aim of speeding up China's progress in achieving the SDGs and contributing to the global implementation of the SDGs.

I. Enhancing the consistency of cross-regional and cross-sectoral governance policies

Establishing a cross-regional coordination mechanism to avoid conflicts in development goals across regions. Through regular joint meetings and policy dialogues, strengthening information sharing and policy coordination to ensure the consistency of policy goals in different regions. However, diversity in implementation methods should be allowed to maximize the overall effectiveness. Meanwhile, cross-departmental collaboration should be strengthened, and a closer policy coordination mechanism should be established. Building an efficient intra-governmental collaboration platform can help improve the coordination of multi-departmental policy goals and ensure the balanced implementation of SDGs nationwide.

II. Leveraging synergies, managing trade-offs, and focusing on priority goals

By evaluating the impact intensity of each goal within synergy and trade-off networks, key goals critical to achieving the 17 SDGs can be identified and prioritized. Research suggests that China should prioritize SDG 7 (clean energy), SDG 12 (responsible consumption and production), SDG 14 and SDG 15 (ecosystem protection), and SDG 8 and SDG 10 (social equity and reduced inequalities). However, priority goals should be adjusted according to different development stages and regional characteristics to maintain consistency between local and overall goals, as well as between short- and long- term goals.

III. Strengthening basic theoretical research for the implementation of SDGs

Establishing special research funds to support SDGs-related research in universities, research institutes, and enterprises. Encouraging interdisciplinary research to explore key scientific questions in the SDGs field. Furthermore, setting up SDGs achievement policy research centers to gather experts in theory, modeling, and policy research. These centers can regularly release research reports and offer practical recommendations to directly inform policy-making and adjustment.

IV. Strengthening data acquisition capabilities to fill the SDGs data gap

Establishing a data monitoring and management system covering multiple fields such as social, economic, and environmental. Ensuring the timeliness and accuracy of data through technical means to provide comprehensive data support for the implementation of SDGs. Encouraging the government to establish data partnership relationships with enterprises and social organizations to fill the SDGs data gap through data resource sharing. Establishing a reward mechanism to stimulate active participation in data sharing, improving the utilization benefits of data, and providing a more reliable data foundation for the implementation of SDGs.

Originally published in Chinese in *Bulletin of Chinese Academy of Sciences*, Issue 5, 2024

Authors: Fu Bojie, Zhang Junze

Pathways to Accelerate the Achievement of Sustainable Development Goals

The United Nations Sustainable Development Goals (SDGs) align closely with the vision of building a human community with a shared future, sharing a common aspiration to strengthen global governance. Major initiatives such as targeted poverty alleviation, ecological civilization construction, and the Belt and Road Initiative are integral processes for advancing this shared future, and they simultaneously contribute to accelerating the implementation of the SDGs.

Achieving the SDGs holistically remains a critical challenge. Given the urgency of these goals, the complexity of their interlinkages, and the disparities in national development levels and scale effects, exploring universally adaptable pathways to implement the SDGs is essential for global prosperity and stability.

By summarizing the common characteristics of actions taken in recent years to build a human community with a shared future, we propose the "Classification–Coordination–Collaboration" framework to accelerate the global achievement of the SDGs.

Classification serves as the foundation, aiming to identify the principal characteristics of different SDGs, disparities in national development, and cross-scale impacts. Coordination is the core component, ensuring policy coherence, enhancing interdepartmental and intergovernmental synergy, and strengthening global connectivity. Collaboration is the effective means for holistic implementation, emphasizing partnerships in economic, technological, and cultural domains.

Classification

Classification involves grouping entities based on their attributes, hierarchies, or other features, laying the groundwork for analyzing relationships, comparing differences, and enabling joint governance.

Existing studies often categorize SDGs using the traditional "social, environmental, and economic" framework, which fails to fully capture their systemic interconnections. Therefore, a systematic approach is needed to clarify the logical relationships among SDGs and classify them accordingly.

At the national scale, countries should be classified based on socioeconomic development disparities to identify their strengths and weaknesses in implementing SDGs, thereby fostering joint actions. From a spatiotemporal perspective, SDGs are global-scale objectives. National actions must account for cross-border externalities to avoid adverse impacts on other countries, while leveraging domestic advantages to prioritize short-term breakthroughs and long-term progress.

Coordination

Coordination is a management approach that requires top-level design support and unifies the exercise of power and responsibility. This implies the need for command and operational capabilities.

The purpose of coordination is to solve the problem of unbalanced development. It helps to reduce social management costs, improve coordination and policy consistency between different management departments and different levels of government, and promote social equality in general.

In the process of achieving SDGs as a whole, coordinated management needs to consider the interrelationships between different categories of SDGs, the capabilities and needs of different countries and regions, and the coordination between different policies, including: ① Coordination among different categories of goals; ② Coordination among different countries and regions; ③ Policy coordination across scales.

Collaboration

Sustainable development is inherently a product of multi-sectoral cooperation, making collaboration an imperative. Building on the above analysis, international and regional collaboration should be strengthened in three domains: ① Economic, enhancing global economic partnerships; ② Technological, promoting innovation and knowledge-sharing; ③ Cultural, cultivating a global culture of sustainability. These areas are complementary, all taking into account the complex relationships between the SDGs, the needs and constraints of different countries, and the scale effects of joint actions. Strengthening collaboration in any of these areas can have a positive impact on all the SDGs. However, this collaborative relationship depends on people's choices and can be managed. If people choose to work together and govern collaboratively, it will have significant implications for overall achieving the SDGs.

The "Classification–Coordination–Collaboration" framework incorporates principles from initiatives like the Belt and Road Initiative and ecological civilization construction, supported by numerous successful cases. Overall, this framework aims to respect differences and promote the participation of all countries in global governance, ensuring short-term breakthroughs and long-term overall progress in achieving the SDGs. With only about 11 years remaining until 2030, countries around the world should join efforts to address the challenges of implementing the SDGs by strengthening collaboration in economic, technological, and cultural fields. This framework represents an important exploration in promoting the global achievement of the SDGs.

Originally published in Chinese in *China Sustainability Tribune*, Issue Z2, 2019

Authors: Fu Bojie, Wang Shuai, Zhang Junze

A Modern Governance System for Global Sustainable Development Goals

Background

In September 2015, member states of the United Nations reached consensus on the 2030 Agenda for Sustainable Development. This agenda outlines 17 Sustainable Development Goals (SDGs) and 169 targets, reflecting the global commitment of the international community to achieving sustainable development across social, economic, and environmental dimensions. The uniqueness of the SDGs lies in their broad international acceptance, the universal pledge to their implementation, and the integrated, indivisible understanding of the interlinkages among all 17 goals. While the SDGs provide a framework for collective action against global challenges, the 2024 Sustainable Development Goals Report reveals that only 17% of the targets are on track, nearly half show limited or moderate progress, and over one-third have stalled or regressed.

The sluggish progress of the SDGs has sparked extensive academic debate and attention. UNESCO designated 2022 as the "International Year of Basic Sciences for Sustainable Development". In August 2023, the United Nations General Assembly adopted the resolution International Decade of Sciences for Sustainable Development 2024–2033, to be spearheaded by UNESCO. Notably, in September 2024, the UN Futures Summit in New York adopted the Pact for the Future and its annexes, the Global Digital Compact and the Declaration on Future Generations. The Pact for the Future encompasses five key areas—sustainable development and financing, international peace and security, science, technology, innovation and digital cooperation, youth and future generations, and transforming global governance—with 56 specific commitment actions. These efforts underscore the international community's heightened focus on addressing sustainability challenges.

However, the key to leveraging science for SDG achievement lies in clarifying the interrelationships among the goals, particularly their synergies and trade-offs. Existing research indicates that the complex interplay among the 17 SDGs demands comprehensive, integrated, and consistent decision-making. Yet, decision-making, implementation, and monitoring of the SDGs involve participation from the public and private sectors, as well as civil society. Regional disparities and conflicts, along with uncertainties of global change, often increase this complexity. Scholars have highlighted that SDG governance requires creating an environment conducive to collective action, ensuring accountability of relevant stakeholders, and addressing the complex interrelationships among the goals. Governance is increasingly seen as the fourth pillar, following social, economic, and environmental aspects. There is an urgent need for a scientific framework to clarify the complex interrelationships and dynamic mechanisms among the 17 SDGs.

Integrating scientific discoveries with governance policies is essential for accelerating SDG achievement. To this end, interdisciplinary teams of natural and social scientists must first analyze the interrelationships of the 17 SDGs within multi-level complex networks (global, regional, national, and local). Second, the scientific community needs to identify the dynamic mechanisms of multi-level complexity governance and the conditions for system stability. Based on these analyses, policymakers can establish effective governance systems to optimize policy-making and advance SDG implementation.

Given the current lack of in-depth research and discussion on SDG governance systems, these tasks remain highly challenging. This paper aims to further explore SDG governance systems based on existing understanding of SDG interrelationships, providing a theoretical basis for improving SDG governance models and promoting the development of modern governance systems and capabilities for SDG achievement.

System for Implementing SDGs Governance

I. A systems classification framework for the SDGs

Traditional studies categorize the 17 SDGs into social, economic, and environmental domains. However, at the level of specific targets and indicators, any single goal may encompass metrics from all three categories. Thus, simplistically labeling a goal as purely social, economic, or environmental risks oversimplification and may lead to fragmented implementation approaches. From a systemic perspective, the 17 SDGs can be classified into three groups based on their inherent attributes: "Essential needs", "Objectives", and "Governance".

"Essential needs" are about human survival and include SDG2 (Zero Hunger), SDG6 (Clean Water and Sanitation), SDG7 (Affordable and Clean Energy), SDG14 (Life Below Water), and SDG15 (Life on Land). "Objectives" represent aspirations for material and spiritual well-being beyond basic survival. They cover SDG1 (No Poverty), SDG3 (Good Health and Well-being), SDG4 (Quality Education), SDG5 (Gender Equality), SDG8 (Decent Work and Economic Growth), SDG10 (Reduced Inequalities), and SDG16 (Peace, Justice and Strong Institutions). "Governance" involves key coordinating actions and includes SDG9 (Industry, Innovation and Infrastructure), SDG11 (Sustainable Cities and Communities), SDG12 (Responsible Consumption and Production), SDG13 (Climate Action), and SDG17 (Partnerships for the Goals).

Under this framework, achieving the SDGs entails minimizing resource consumption for "Essential needs" (i.e., resource-efficient utilization) and maximizing the fulfillment of "Objectives". This balance relies on "Governance" group SDGs to harmonize competing priorities across "Essential needs" and "Objectives".

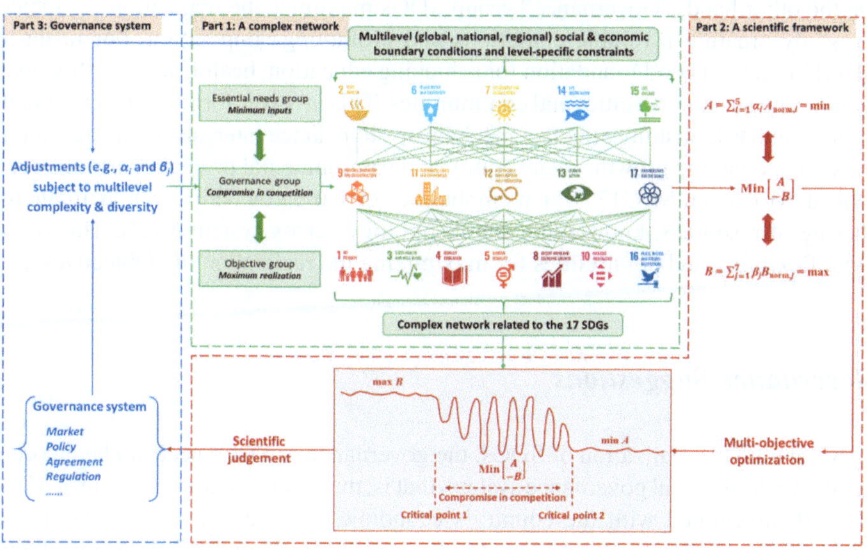

Scientific analysis framework and governance system for the multi-level complexity of Sustainable Development Goals

II. Coordinating mechanisms for the "governance" group SDGs

The 17 SDGs form a multi-level system (global, regional, national, and local). Differences across levels manifest as changes in boundary conditions and internal structures, while commonalities emerge in the principle of compromise-in-competition during interactions among the SDGs. Transforming this into a multi-objective optimization problem creates a "three-zone and two-tipping-point" framework. It ensures common scientific principles for sustainability across levels. "Essential needs" focus on reducing resource waste and boosting resource efficiency. They benefit more from progress in natural sciences and technology. In contrast, achieving "Objectives" emphasizes changing distribution patterns of goods and services and reducing inequality. It relies more on institutional design, which involves social sciences, ethics, and morals. The main function of "Governance" group SDGs is to balance "Essential needs" and "Objectives". They ensure efficient resource allocation and utilization and drive overall SDG progress.

On the one hand, "Governance" group SDGs minimize resource-use for "Essential needs" by promoting innovation, optimizing infrastructure, and transforming production and consumption. For example, SDG9 and SDG12 can enhance resource efficiency and reduce natural resource consumption through technological innovation and sustainable production. Meanwhile, SDG11 and SDG13 can cut environmental pressure and boost resource efficiency by improving urban planning and addressing climate change. This helps guarantee food (SDG2) and water (SDG6) security and strengthens the protection of aquatic (SDG14) and terrestrial (SDG15) life systems. On the other hand, "Governance" group SDGs maximize the realization of "Objectives" by creating favorable conditions and promoting cooperation. For instance, SDG11 creates a good foundation for achieving education, health, and equality goals by building sustainable cities and communities. This involves formulating reasonable policies and establishing effective mechanisms to balance interests. It ensures that in pursuing economic growth, social equity and environmental protection are not overlooked. Moreover, SDG17, which emphasizes global partnerships, can facilitate the sharing of resources, knowledge, and technology across system levels. This offers more flexibility and possibilities for maximizing the realization of "Objectives".

Governance Suggestions

To accelerate the realization of SDGs, the governance of SDGs needs to be incorporated into the national governance system, that is, the specific actions to achieve SDGs are fully integrated with the country's economic governance, social governance, ecological governance, cultural governance, and political governance. Among them, economic governance needs to promote green transformation, encourage efficient use of resources and environmental protection through policy incentives and market mechanisms; social governance should focus on improving public welfare, ensuring that education, medical care and other public infrastructure can cover all members of society, especially vulnerable groups, to promote social fairness and inclusive growth; cultural governance is about promoting the cultural values of sustainable development, promoting widespread recognition and acceptance of SDGs through education and media publicity; ecological governance should focus on protecting and restoring ecosystems to enhance biodiversity and ecosystem functions; political governance plays a core role in ensuring policy consistency and systematization among different departments, regions, and levels of government, which requires cross-departmental, cross-regional, and cross-level government coordination and close cooperation, thus providing institutional guarantees for accelerating the realization of SDGs.

"When the ecology thrives, civilization thrives; when ecology declines, civilization declines" is General Secretary Xi Jinping's precise judgment on the harmonious coexistence relationship between humans and nature. Strengthening ecological governance is crucial for maintaining the Earth's life support system. This process requires a more comprehensive and systematic consideration of resource management, environmental management, and ecosystem management under a scientific framework to intervene and prevent the occurrence of risks. Resource management aims to allocate and efficiently use natural resources, such as water resources, land, energy, and minerals. Sustainable use of resources helps to promote the green transformation of the economy and ensure long-term social welfare. Environmental management focuses on reducing pollution, protecting air quality, water quality, and soil health, and responding to climate change. Implementing strict environmental protection regulations and promoting clean technology are necessary for environmental management. Ecosystem management targets various types of ecosystems, such as forests, grasslands, wetlands, and marine ecosystems, focusing on protecting ecosystem integrity and systematically enhancing ecosystem functions. Carrying out in-depth ecological protection and restoration projects for mountains, rivers, forests, farmlands, lakes, and grasslands will provide human society with more stable and high-quality ecological products, while demonstrating China's sense of responsibility and leadership in global ecological governance.

However, how to establish the complex network system shown in the figure above deserves the full attention of decision-makers at all levels, especially how to further analyze the dynamic mechanisms and system stability conditions in the complex system corresponding to this framework. In order to achieve the goal of scientifically managing and regulating such systems, the formation of cross-scientific teams at different levels is needed to conduct in-depth systematic research on specific cases based on the principles of complexity at different levels, and to provide support for the scientific management of socio–economic–ecological systems.

Conclusion

A modern governance system facing the global SDGs requires a new kind of systemic thinking. Classifying the 17 SDGs into "Essential needs", "Objectives", and "Governance" clarifies their interrelationships and interaction mechanisms, which is key to governance capacity. The "Governance" group goals play a vital coordinating role by minimizing "Essential needs" consumption through innovation and efficiency while maximizing "Objectives" achievement by creating favorable conditions. This framework offers a new perspective on SDG implementation and highlights the importance of combining scientific analysis with policy-making. Without a sound scientific framework, these governance goals cannot function effectively.

In the process of implementation, SDG governance must be integrated into national governance systems, covering economic, social, cultural, ecological, and political dimensions. Economic governance should drive green transitions; social governance should enhance public welfare; cultural governance should promote sustainable development values; ecological governance should strengthen sustainable resource use, environmental protection, and ecosystem conservation and restoration; and political governance should ensure policy consistency and promote coordination and cooperation across department, region, and government levels. This systemic approach will help enhance SDG governance modernization and provide important examples for global sustainable development.

Looking to the Future: Aim High with the Soul of Geography

Speech at the 2019 Graduation Ceremony of the Faculty of Geographical Science, Beijing Normal University

Dear students, teachers, and parents:

Good afternoon!

Today, we are holding the grand graduation ceremony of the 2019 class of the Faculty of Geographical Science, Beijing Normal University. On behalf of all faculty and students of the Faculty of Geographical Science, I would like to once again congratulate all graduates of the Class of 2019 on successfully completing your studies and graduating smoothly! Congratulations on the new chapter of life you are about to begin!

On this day, which is worth remembering for a lifetime, we want to sincerely thank the parents, teachers, and fellow students who have quietly contributed, selflessly dedicated themselves, and cheered you on throughout your growth and development. Let us all pay tribute to them and give them the warmest round of applause!

Dear students, the time you spent studying at university is a precious memory in life and a brilliant movement of youth. I feel extremely gratified that the Faculty of Geographical Science has been able to accompany and witness your growth. At the inception of our Faculty in 2016, we clearly set our construction goals. While building a Chinese school of geography recognized internationally, we also prioritized student training, aiming to establish a model for higher geography education. We strive to cultivate research talents in geography and related disciplines with potential for development, as well as to train high-quality, comprehensive, and applied talents with geographic thinking. At the same time, we actively train excellent geography teachers who are professionally solid and deeply care for their students.

Over the past few years, I have seen you lead fulfilling academic lives—seriously studying professional courses, actively exploring your personal learning and research interests, and conducting solid scientific research. I have seen more and more students produce outstanding academic achievements during their studies, with some being published in top journals of related disciplines and gaining recognition and attention from peers. I've also heard students say that, by taking advantage of the Faculty's growing number of international exchange platforms, they've broadened their horizons and expanded their personal development directions. You have also paid attention to comprehensive development, cared about the motherland and society, participated in social practice and teaching support activities all over the country during winter and summer vacations, rooted in the people and serving them. You've participated in major national conferences and events, and completed various volunteer services with excellence. In the annual Geography Culture Festival, your enthusiastic participation has spread geographic knowledge and promoted science. In various geography-related academic competitions at home and abroad, your victories have come frequently. You have forged ahead in academic research, grown your talents through campus activities, served society through volunteer work, and displayed your skills in innovation and entrepreneurship. Every effort of yours has contributed to a more mature and better self. The Faculty of Geographical Science is proud of you! The Faculty of Geographical Science applauds you!

Since the founding of the Faculty, I have spoken with you many times about the development and applications of geography. Today, as I bid you farewell, I'd like to speak about geography and the individual.

I believe that through several years of learning and practice, you have not only acquired knowledge of geography but also developed geographic thinking. Geography values practice, and a spirit of hard work and down-to-earth practice provides the foundation of our profession. From the first to the third year, our undergraduates go on fieldwork almost every summer. You observe geological structures, identify plants, understand ecosystems, and enter cities to observe the spatial patterns of human activities. At the graduate level, fieldwork becomes significantly more difficult and demanding. Some of you have spent extended periods in the field—enduring wind, digging soil, climbing mountains, and crossing seas. Others have conducted surveys in remote regions, living and eating with local farmers and herders. Though fieldwork is grueling and filled with unforeseen challenges, it allows us to obtain firsthand research data and grasp authentic patterns. It is the fundamental basis for generating original research. After graduation, whether you continue in geography-related scientific research or move into other fields, I hope you will continue to carry forward the spirit of enduring hardship and staying grounded—let go of past honors and achievements, lower yourself, start from small things, start from originality, seek truth and be pragmatic, and lay a solid foundation for personal growth.

Geography emphasizes scale, and the grand vision of transcending limitations and looking to the future determines the height of development. In geography, we often discuss spatial and temporal scales—from a leaf, a tree, a forest, to local, regional, and global levels. Studying problems at different scales leads to different methods and discovered patterns. For instance, in studying global climate change, research over the past few decades may reveal a pattern of decreasing temperatures year by year. But when the time scale is extended, those decades might turn out to be a phase of overall historical temperature rise, making future patterns one of warming. In your future work and life, you will also encounter similar principles. If you only focus on immediate interests, you may become fixated on temporary gains and losses. But if you choose based on collective interest and long-term development, you can let go of ego and take on more responsibilities, achieving greater undertakings. I hope every geography student who leaves here will set lofty ideals, always maintain a sense of devotion to the country, be willing to contribute, dare to take responsibility, and become outstanding talents who can accomplish great things.

Geography requires integration, and a comprehensive, thoughtful, inclusive spirit of harmony leads to a fulfilling life. Geography takes the most complex system of nature and humans as its research object, involving various disciplines and requiring support from multiple fields of knowledge and methods. Last year, at the opening ceremony for freshman, I said that in the new era, geography is characterized by interdisciplinarity and integration, and it emphasizes team spirit. I hope our graduates will apply this integrative geographic thinking to future work and life—whether facing a work team or work objects—by understanding and tapping into each person's strengths, accepting differences, and embracing diversity. "Harmony" is the essence of traditional Chinese culture. By implementing the concept of harmonious coexistence into action, achieving balance and harmony, we can make our lives more peaceful and happier.

Dear students, in a few days, Beijing Normal University will officially become your alma mater. Please remember that here are teachers who will always care for you, and every blade of grass and tree that witnessed your growth. I hope you'll come back often—return in times of success to share joy and pass on your experiences to your teachers and classmates; return in times of confusion to walk quietly around campus and reconnect with your original aspirations; return during your struggles and progress to revisit the classroom—geographic knowledge may continue to inspire and support you.

Finally, I wish you all a bright future, peace, and happiness!

Thank you all!

Postscript

Li Xiubin

There is a peculiar phenomenon in the geography community: communication among peers is less frequent than with scholars from other disciplines. For instance, even within the same geography department, there are few common topics between faculty members teaching physical geography and those teaching human geography. However, they often maintain close contact with scholars from other disciplines who share the same or similar research subjects.

In Prof. Fu's view, it is due to disciplinary differentiation that geography is increasingly showing a trend of becoming hollowed out. The discipline's tradition of integrative thinking is gradually fading. Yet, integration—alongside regionality and interdisciplinarity—is an essential characteristic of geography. The future of this ancient discipline equally depends on the implementation of integrative thinking.

The integrative thinking of traditional geography stems from an understanding of the complexity of its research objects, emphasizing the integration of elements. Natural elements such as geology, landforms, hydrology, soil, and vegetation; socioeconomic elements such as population, location, settlements, and transportation. To deeply understand a place or explain a phenomenon, one must comprehensively grasp various elements and thoroughly analyze their interconnections and interactions—only then can one achieve profound insight. So where does integration manifest in modern geography?

To this end, Prof. Fu has constructed a set of theories and methods for integrative geographic research. Beyond traditional element integration, he particularly emphasizes two aspects: one is the coupling of patterns and processes; the other is data integration and model development. The research subject revolves around the coupled human and natural system from an integrated perspective.

Pattern refers to geographic spatial distribution; process refers to spatiotemporal evolution; coupling describes the feedback relationship in which pattern influences process and process alters pattern. The coupling of patterns and processes is mainly achieved through two approaches: direct observation and system simulation. Direct observation is suitable for smaller spatial scales, while system analysis and simulation are appropriate for larger spatial scales. The coupling of patterns and processes has become an important approach to studying natural process mechanisms such as

B. Fu, *The Soul of Geography*, https://doi.org/10.1007/978-981-95-1062-7

the carbon–water cycle, ecohydrological processes, and soil–vegetation–atmosphere interactions. It is also a vital method for examining socioeconomic processes such as population migration, transportation and logistics, and urban–rural integration, playing a key role in the comprehensive understanding and prediction of geographical phenomena.

Data integration and model development are most clearly reflected in the simulation of the coupled human and natural system. Guided by systems thinking and based on research into elements, patterns, and processes, coupled simulation quantitatively characterizes the complex internal and external interactions within the human–natural system. Such simulation not only provides the foundation for understanding the feedback relationships, dynamic evolution, and overall functions among the components of this complex mega-system, but also serves as a tool for optimizing and regulating human–nature relationships. Simulations and experiments on system states and processes using this tool can provide decision support for regional or global sustainable development. It can be said that the progression from element integration to system integration represents a step toward the quantification of integrative geographic research.

In fact, not only the geography community but also the recently emerging fields of Earth system science and sustainability science greatly value integrative thinking and are exploring approaches and methods for integrative research. I have observed that, over the past thirty years with the emergence of concepts such as virtual water and ecological footprint, the focus of human–nature relationship research has shifted from a long-standing concern with environmental carrying capacity to a new perspective centered on human consumption behavior. Various consumption indicators such as water, land, energy, and carbon have been designed, enabling the simultaneous assessment of the multiple efficiencies and impacts of a single human activity. This has enhanced the multidimensional and integrated understanding of a phenomenon. Such integration might be called "polyhedral integration."

There is another kind of integration, which, following this naming style, could be referred to as "jigsaw integration". As Prof. Fu describes the third type, it entails breaking a large system into several modules. Each discipline or team is responsible for just one of them. The underlying assumption is that "the mission of each discipline and team is limited", and each team focuses only on its core module of responsibility. This essentially accepts the reality of disciplinary specialization and then proceeds to reassemble: your model's output becomes the input for mine. Some argue that in today's era of deep disciplinary differentiation, no single discipline can encompass everything, and that if one aims to conduct integrative research on complex phenomena, this jigsaw-style approach is the only viable option.

For Prof. Fu, the notion that "integration is the soul of geography" is not empty talk—it is a principle deeply practiced throughout his scientific career. Fu's team, based in Yangjuangou, has conducted long-term field observations and experiments for decades. Their revelations of how land use types and their spatial configurations in this catchment affect soil moisture, nutrients, and erosion exemplify the coupling of land use patterns and ecological processes that Fu emphasized, and represent a new path for integrative geographic research.

From Yangjuangou to the entire Loess Plateau, to the Yellow River Basin, and onward to the national and even global scale; from land use types and evaluation, to landscape patterns and ecological processes, then to ecosystem processes and services, and ultimately to coupled human and natural systems and sustainable development—Prof. Fu has led his team in applying the integrative thinking of geography to problem-oriented research that serves society, continuously exploring the approaches and methods of integrative research.

The greatest challenge in geographic research lies in how to integrate—this is a consensus among many geographers. Fu's success in this regard lies in his willingness to confront real-world problems and explore solutions through practice. This also echoes two well-known sayings in the Chinese geography community: one is "Geography is a discipline meant to serve practical governance", and the other is "Missions lead the discipline forward".